DATE DUE

ILL SEP 0	8 2006		

Demco

UNCOMMON SENSE

What's wrong with the painting at the inn? (Photo by John Hubbard)

UNCOMMON SENSE

Understanding Nature's Truths Across Time and Culture

ANTHONY AVENI

UNIVERSITY PRESS OF COLORADO

Published by the University Press of Colorado
5589 Arapahoe Avenue, Suite 206C
Boulder, Colorado 80303

Printed in the United States of America

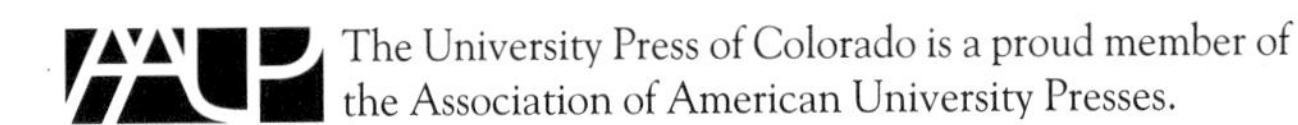
The University Press of Colorado is a proud member of
the Association of American University Presses.

The University Press of Colorado is a cooperative publishing enterprise supported, in part, by Adams State College, Colorado State University, Fort Lewis College, Mesa State College, Metropolitan State College of Denver, University of Colorado, University of Northern Colorado, and Western State College of Colorado.

∞ The paper used in this publication meets the minimum requirements of the American National Standard for Information Sciences—Permanence of Paper for Printed Library Materials. ANSI Z39.48-1992

Library of Congress Cataloging-in-Publication Data

Aveni, Anthony F.
Uncommon sense : understanding nature's truths across time and culture / Anthony Aveni.
p. cm.
Includes bibliographical references and index.
ISBN-13: 978-0-87081-828-8 (alk. paper)
ISBN-10: 0-87081-828-7 (alk. paper)
1. Astronomy. 2. Astronomy—Social aspects. 3. Astrology. 4. Space and time. 5. Science—Social aspects. 6. Science—Philosophy. I. Title.
QB43.3.A94 2006
520—dc22

2005037323

Design by Daniel Pratt

15 14 13 12 11 10 09 08 07 06 10 9 8 7 6 5 4 3 2 1

To Faith Hamlin,
first among friends, second to none among supporters.

CONTENTS

ILLUSTRATIONS

PREFACE

There is a large painting over the dining room fireplace in the inn that marks the center of the tiny village where I live (frontispiece). Painted by a mid-twentieth-century local artist, it depicts what he imagined our town looked like about a century and a half ago. Men in top hats and high boots walk the streets in front of the inn, then dignified by tall, white pillars flanking its entryways. From the balcony a woman overlooks a horse and buggy about to exit at the right of the scene. She is clad in a long dress with a bustle. (How did they keep them clean as they dragged in the thick coating of mud that layered our unpaved streets then?) Sharp afternoon shadows fall on the rooflines from left to right across the chimneys, lending stark contrast to the pale blue daylight sky. A thin crescent moon shines over the tops of the buildings, the lunar cusps pointing upward and to the right—the way you might expect to see the waxing moon on a late afternoon in early summer. Details familiar to me suggest that the painter faced northwest. A walk around the inn convinces me that he had positioned his easel on the northeast corner of the street opposite the building, next to the village green and just in front of the village office, confirming my impression.

"What's wrong with the picture at the inn?" That was the question I posed some years ago as a young teacher of astronomy bent on challenging his students to extracurricular academic heights—"ten points on the next exam to anyone who can figure it out!" Over the years dozens of achievers curious enough to hike the half-mile down the hill that harbors our lofty house of learning to visit the inn have failed to solve the puzzle. The handful who did were paid off handsomely.

What's "wrong" is this: anyone familiar with the sky knows that wherever you see a thin crescent moon in the sky, the sun can't be far away. If the crescent moon were a bow, its arrow must be aimed at the sun. To find it, connect a line between the horns of the crescent and extend a perpendicular from its midpoint through the lighted portion of the moon. The thinner the crescent, the closer you'll find the sun along that line. In the case of the village scene, since the cusps point slightly upward and to the right, the sun must lie downward and to the left, very close to the northwest horizon (evidently obscured by the building's façade). But then, why are all the shadows in the scene directed as if lit by a sun shining over the left shoulder of the viewer? To make matters worse, the tilt of the cusps (pointing to the two o'clock position) is what you'd expect for a location in a latitude somewhere south of the equator. Late afternoon waxing crescent moons tilt the other way (toward the ten o'clock position) in northern hemisphere mid-latitudes.

Did the artist add a moon to the scene as an afterthought? Perhaps he remembered that it was positioned there at a different time of the day and year than when he made the painting. Or maybe when he returned later to finish the scene in which he had already painted in the shadows, he sketched in the moon as he saw it at that time. There are numerous other possibilities to explain why the artist's representation of nature is out of whack with reality, but the point is that he got it wrong.

I remember the expressions of delight, satisfaction, and maybe even a touch of smugness on the faces of my students who were able to solve the mystery at the inn (and perhaps even got the folks to spring for a decent meal as recompense for having neglected the precious offspring they sent far away to university). The artist's faux pas has become something of a curricular legend over the forty-odd years my kids have been dragging their parents down to tables beside the fireplace so that they might brag about their intimate knowledge of the firmament and their membership in the elite fraternity of scientific decoders of reality-flawed art.

At the time I first posed the "inn problem," I didn't realize that I was tapping into the nascent vein of what would become a healthy cottage industry—scientific sleuths discovering the naked truth that underlies famous works of art. Several of Vincent Van Gogh's works have been subjected to this sort of scientific exercise in demystification. Van Gogh liked to paint background scenes with stars amidst swirling clouds. His *Starry Night Over the Rhone* (1889), the inspiration for Don McLean's popular song "Starry Starry Night," depicts a clear replication of the stars of the Big Dipper. His diaries reveal frequent comments about the cosmic scenery: "This morning I saw the country from my window a long time before sunrise, with nothing but the morning star, which looked very big," he once wrote his brother Theo.[1] In fact, the bright white object just above the horizon in Van Gogh's *Starry Night,* according to astronomically oriented art detective Donald Olson, must be Venus. He proved it by calculating the exact time and place that fit with the situation of the crescent moon at the right end of the scene. The same cusp-perpendicular rule applies because the planets, the sun, and the moon essentially follow the same track—the ecliptic, which is marked by the constellations of the zodiac.

Cosmic gumshoes also fixed *White House at Night* in real time. Van Gogh painted it on June 16, 1890, at 7:00 PM! Once they computed exactly when Venus was perched just over the two-story, tile-roofed house at the center of the frame, astronomers traveled to Auvers-sur-Oise, just northwest of Paris, to locate the precise place that matched the exact time. One of them exclaimed that he was astonished by the accuracy of the position of Venus in the picture. Incidentally, six weeks after he painted *White House,* Van Gogh committed suicide.

A NASA-trained team of scientists recently published their findings on the real truth about Norwegian painter Edvard Munch's most famous work,

The Scream (Figure 1). Previously thought to have been composed in 1893, although several versions were completed sometime between 1892 and 1896, this late Impressionistic piece prominently features a horrified-looking genderless figure clad in black.[2] The hollow eyes, wide-open mouth, and hands clasped over the ears render the scream almost audible to the viewer. The mood is intensified by confused swirlings of indigo water and a bright yellow-red twilit sky in the background's harbor scene behind the walkway and railing on which the figure stands.

In January 1892 Munch offered a written description of an event that seemed to match what he had transferred to canvas:

> I was walking along the road with two friends—then the sun set—all at once the sky became blood red—and I felt overcome with melancholy. I stood still and leaned against the railing, dead tired—clouds like blood and tongues of fire above the blue-black fjord and the city. My friends went on, and I stood alone, trembling with anxiety. I felt a great, unending scream piercing through nature."[3]

Experts from the art world have remarked that red and yellow clouds are a climatic oddity in northern Europe but often appear in paintings from that area. Others think the swirls are visualizations of sound waves or perhaps externalized versions of pent-up energy. But the NASA researchers were not satisfied with internal, psychological explanations; they were looking for concrete, external stimuli to explain the background of the painting. They posed a specific set of questions: Exactly what did Munch see in the sky? In what specific direction was he viewing? And precisely when did he take that memorable walk?

The "scream piercing through nature," they determined, was the eruption of Krakatoa, a volcano in Indonesia that blew up in an instant on August 11, 1883. Not only did Krakatoa produce sounds heard a thousand miles away—the cataclysm was recorded on seismographs all over the world twice over—but Krakatoa was also responsible for some of the most dramatic scarlet and crimson sunsets ever witnessed. Three months after the explosion the *New York Times* reported startled people in the streets gazing at the "bloody red" sky. Many of them believed there must be a great fire in the distance. Extraordinary twilights were reported worldwide up to a year after the blast.

1. Edvard Munch's The Scream *may hold clues telling precisely where and when it was painted. (©2006 The Munch Museum/The Munch-Ellingsen Group/Artist Rights Society (ARS), New York)*

But if Munch wrote his description of the scream in 1892—and if he painted *The Scream* around that time—how can a red sky resulting from a volcanic eruption on the other side of the world nine years earlier have been the root cause? "Science can still explain the Krakatoa-induced sky," argue the NASA researchers.[4] Their team has analyzed Munch's correspondence with friends, which indicated that the "scream event" took place considerably earlier. In fact, several sketches survive from as early as 1885. With single-minded persistence, the investigators gleaned clues from the artist's diaries. They ferreted out vital information that led them to the exact spot (a wooded hill in western Oslo overlooking the bay) where—they insist—Munch was overcome by the volcanically induced sunset.

Reading my account of the adventures of scientific sleuths in the world of art you may wonder—as I did when I chose to use it in my introduction—why is there such a deep and persistent interest in pulling away the veil that shelters the naked truth about exactly where Van Gogh and Munch were and specifically what they saw and precisely when they saw it? Knowing exactly where and precisely when seem like harmless inquiries. They don't conflict with the deeper issue of trying to sense what the artists felt when they created their masterworks. But most of us are not as interested in an artist's feelings as we are in decoding the mystery. My anecdotes about reality-flawed works are intended to highlight one of contemporary culture's preoccupations—finding literal truth.

Why this fascination with literal truth? Why the deep desire to demystify a mystery? Is it human nature, or are we simply conditioned by attitudes peculiar to the age we live in? Have people always felt this way? People of *all* cultures? I'm not about to offer definitive answers to these vexing questions. Nor am I out to decode any more puzzles. What I'm after is something more fundamental—that what we think truth is and how we have gone about seeking it through the ages sheds considerable light on how we operate as rational beings, on how we pose questions, and on why the questions we think are important always seem to change or get rephrased. I am astonished by the diversity of approaches to acquiring these truths—how "what really matters" is altered through time.

My subject matter is the natural world and what we think is worth knowing about it. In Part One I deal with several different "common sense" approaches to finding truth in nature. Such approaches emerge out of different questions posed by people with different backgrounds and interests. The value of knowledge depends on who seeks it. For example, I explore and contrast questions that have been asked over the millennia about the Star of Bethlehem. There's the scientist's question, what was it? The historian's question, was there a star? And the theologian's question, what did it mean? If we are not exposed to the questions someone else asks, we might develop a tendency to think we're the only ones who are asking the right questions. I also deal with the question, how is the universe arranged? I trace the origin of our contemporary common sense that understands heaven as a hierarchy of motion on orbits governed by natural laws. I contrast that with the hierarchy of layers patterned after the structure of society that emerges from the common sense of so many other cultures of the world. Other chapters in Part One entertain questions about the age of the earth, the development of life, and the death of the dinosaurs. I am less concerned about right and wrong answers. My focus is on the uncommon senses human culture exhibits when confronted by phenomena in the natural world. By exploring conflicting points of view, I demonstrate that scientific discourse is, at least in part, a social enterprise conditioned by changing values.

There once was a game played in ancient Mesoamerica called *patolli.* It consisted of a board shaped like a cross, each of its four arms divided lengthwise and segmented into small rectangular spaces about which opponents moved "men" in accordance with numbers indicated by the roll of dice. If you landed on a space occupied by your opponent's man, you "killed him"; that is, you knocked him off the board and he had to start over. The first one whose men reached home won the game. There's an almost identical game from India known as *pachisi* (*parcheesi*). The theory of cultural diffusion explains the stark similarity between the two games by proposing that someone traveled across a vast sea and brought the game from one place to the other (the east-to-west route is more prominently mentioned); but I think the games were invented independently. For me *patolli* and *parcheesi* are devices that certify a mysterious unity of culture, which is the subject of Part Two. There are certain shared sensibilities that transcend culture. We all make plans in two dimensions that we call maps; we create symbols divided into four parts; we devise analogs to mete out the passage of time; and we all formulate lists so that we can organize and classify the things we see. All of these—space, symbols, numbers, time, and lists—are subjects of chapters that make up Part Two. As I explore each of these shared sensibilities, my intent is to show how, in the human quest for order, they inform shared processes of organizing information about the natural world.

Different times, different people, different circumstances, different questions. All lead to different ideas and concepts that color the point of view we adopt when we try to make sense out of the oft-seeming chaos that characterizes the world around us. So, this book is as much about what we all share as it is about where we differ in arriving at our common sense.

NOTES

1. Quoted in moma.org, Van Gogh Collection (March 18, 2004).

2. Coincidentally the most famous version of *The Scream* was heisted from its home in Norway's Nasjonal Gallereit as I wrote this chapter.

3. Donald Olson, Russell Doescher, and Marilynn Olson, "When the Sky Ran Red," *Sky and Telescope* (February 2004): 28–35; cf. R. Panek, "The Scream, East of Krakatoa," *New York Times* (February 8, 2004), 29.

4. Ibid.

ACKNOWLEDGMENTS

My thanks go out to Cliff Mills and two extraordinarily helpful anonymous readers for comments on earlier drafts of this work, to John Hubbard for his imagery, once again (for the fifth time) to Darrin Pratt and his ever professional staff at the press, especially Laura Furney, Dan Pratt, and Ann Wendland, and to Faith Hamlin, who lives up to her name by consistently seeking value in what I write, not to mention helping to get it into print. Finally, I thank Lorraine Aveni and Diane Janney, who have been just as enduring in helping me to commit my words to paper.

UNCOMMON SENSE

PART ONE

WHERE WE CONFLICT

ONE

WHAT WAS THAT STAR?

We think that the star which appeared in the east . . . is to be classed with the comets which occasionally occur, or meteors, or jar-shaped stars.

—ORIGEN, *CONTRA CELSUM*, AD 248[1]

[M]iracles are less about an event, itself, than the powerful experience of the person who witnesses it.

—R. J. RUSSELL, CENTER FOR THEOLOGY AND THE NATURAL SCIENCES[2]

We live in a world of diminishing imagination space. The need to demystify—the burning curiosity to learn "the real truth" about art that I described above—also pervades the world of literature. What was the Holy Grail? The real Troy? The Emerald Tablet? The Fountain of Youth? The Elixir? The Lost Continent of Atlantis? No single book has been more subjected to literalist probing than the Bible.

This essay was adapted in part from a piece titled "The Star of Bethlehem," which I wrote for *Archaeology* 1998 (November–December): 35–42.

After all, it is a controversial work, and so much is at stake in tracing and reconstructing the belief systems of billions of Judeo-Christians. No shortage of scientifically minded word surgeons have been drawn to dissecting and decoding biblical miracles.

Take the parting of the Red Sea in Exodus 14. Convinced that God rules the earth through the laws of physics, Russian physicist Naum Volzinger has devised a set of mathematical equations to study wave motion induced by wind in a shallow sea. He has concluded that a sustained 24-hour wind averaging 67 miles per hour would have been sufficient to blow the water off the then shallower reef in the northern part of the Gulf of Suez. The fleeing Israelites (not "Israelis" as some of my students call them), believed to have been some 600,000 in number, could have made the shallow crossing on foot in about four hours before the water returned to its normal level.

And what could explain the long-lasting burning bush through which God spoke to Moses when he promised to deliver the Israelites from Egypt? According to British physicist Colin Humphries, the bush (probably a hardy acacia) would not have been consumed if it happened to lay alongside one of the natural gas vents found all over the Sinai. It ignited when Moses brushed against it when walking by. All miracles require good timing!

Biblical mysteries are endless. What did the real Jesus look like? How can biology account for a virgin birth? Did Noah's ark really run aground on Mount Ararat? For decades, "ark-eologists" have searched the environment using clues from Genesis in vain; still they show no sign of giving up. Bruce Feiler, a non-orthodox American Jew has written a best seller, *Walking the Bible,* about his 10,000-mile trek across three continents in search of the Ark, the place where Abraham buried his wife, and the spot where Moses looked out over the Promised Land.[3] He says he wrote the book to find his faith.

Is there a biblical code? Puzzle enthusiasts have spent vast quantities of computer time analyzing equidistant letter sequences. There are more than 300,000 letters in the Hebrew Bible that can be grouped into words and sentences, which many believe to be part of a complicated code, a secret foreknowledge of events that would happen long after the book was written. These events include wars, assassinations, and even the so-called Roswell, New Mexico, flying saucer event. Solve the conundrum and the power of the future awaits you!

As of this writing, the hottest biblical mystery is prompted by Dan Brown's best seller, *The Da Vinci Code.*[4] I have witnessed the success of this book close up; the agency that handles my work also handles Brown's. He outsells me by thousands to one! This novel deals with secret knowledge passed down through the ages to a select few as it follows, allegedly in a fictional mode, the search for the Holy Grail, the lost chalice from which Jesus offered his blood to the disciples at the Last Supper. Its location is a closely guarded secret, passed down through the ages to a select few. The grail turns out to be Mary Magdalene—the "vessel" who held the blood of Jesus when she bore his child. The mystery chronicles the alleged true knowledge of Christ's blood line, which has long been kept secret by clerics. Brown's tale is full of buried truths, codes, and conspiracies, hence its popularity—esoterica gone mainstream.

Now when Jesus was born in Bethlehem of Judea in the days of Herod the king, behold wise men from the East come to Jerusalem, saying "Where is He who has been born King of the Jews? For we have long seen His star in the East, and have come to worship Him."[5]

Thus spoke the apostle Matthew. No Christmas is complete without the telling and retelling of the star story. Every planetarium features its story at year's end. Crèche scenes showing the star over the manger dot village greens, grace store windows, and decorate Christmas cards. The star presides on the pinnacle of practically every Christmas tree (Figure 2). And no matter how many times the story is told, the mystery remains, what exactly was this star that foretold the birth of the Christian savior? There has been no shortage of explanations. Ruth Freitag's bibliography *The Star of Bethlehem* (published by the Library of Congress in 1979) lists 250 major scholarly articles on the subject published in just the first three quarters of the twentieth century. In the 1990s the millennial concern with prophecy spawned several dozen more in addition to at least six full-length books purporting to give *the* answer.

Astronomers have had a field day with the great star. Practically all of them assume without question that it was a real object. One advocate of the astronomical decoding camp best expresses the shared point of view: "In all such cases it seems best to consider as a working hypothesis, that the report is

2. The star on the top of the tree finds its origin in the famous quotation from the Book of Matthew, heralding a star in the east. (Photogram by Lorraine Aveni)

correct, and to investigate whether any astronomical phenomenon exists which fits the report."[6]

Let's run down the short list of key celestial nominees that have graced the stage of scientific explanation since the gnostic philosopher Origen posed the first celestial candidate—a comet, meteor, or jar-shaped star—some two millennia ago (my first epigraph). It simply may have been a bright star, or better, a supernova (a star of extraordinary brightness that blazes forth for a few months), or a recurrent nova (less intense, with several hundred years often intervening between outbursts). Some say the Star of the Messiah was really a constellation of stars, a more likely candidate for a cosmic omen-bearing source than an individual star, or perhaps it was a bright comet (Halley's comet has been mentioned). Others argue that the great luminary was really two comets, a meteor shower, or a fireball (a colossal meteor visible only over a local segment of the earth's atmosphere). The aurora borealis, or northern lights, is another upper atmospheric phenomenon that makes the top twenty list.

Some cosmic detectives prefer to call the Star of Bethlehem the Planet of Bethlehem—bright Venus witnessed hovering over the eastern or western horizon, or transiting the surface of the sun. Combinations of sky phenomena include a close gathering of two or more planets (a conjunction—the one between Jupiter and Venus in 2 BC stands high on the list) or a conjunction plus a comet. The zodiacal light (a reflection of sunlight off interplanetary particles aligned along the ecliptic) has been cited as an explanation, not to mention UFOs—perhaps God's prankish way of announcing good news for humanity. The Urantia movement contends that supermortal aliens channeled Matthew's statement to us. Their gospel states, without equivocation, that Jesus was born on August 21, 7 BC, and that he came to us in a UFO.

Since stars belong in the sky and the sky belongs to astronomers, it should come as no surprise that the professional skywatchers have long been viewed as possessing the keys to unlocking the mystery of the Star of Bethlehem. One December a few years back, I received an impassioned phone call from a Japanese TV reporter from the Tokyo Broadcasting System. He was over the Pacific in a New York–bound 747 in a quest to find the answer to the mystery. At first I could not imagine why Japanese TV would be interested in a program on the Star of Bethlehem, but I quickly figured that he must have read my piece about the star in *Archaeology* and that was the reason for the contact. He pressed me for an interview despite the fact that, to his utter amazement, my university was not in New York City. (Not many foreigners are aware that New York extends several hundred miles beyond the confines of the Big Apple.) When he finally arrived the next day, he and his crew wasted no time converting my tiny office into a mock laboratory, adorning it with all manner of computer terminals and sky globes we hauled out of a basement closet. The clip showed me hovering over a software program pointing to an animated version of a planetary conjunction on the screen as I uttered sound bites about what the star could have been. That done, the journalist opened his wallet, peeled out several $100 bills, and lay them on my desk in recompense for my valuable time. He also bequeathed me a pen and a travel alarm, each engraved with the letters "TBS." I never saw the program but, to judge from my brief experience with the Tokyo Broadcasting System, I would venture a guess that it was all about how scientists decode ancient mysteries.

As evidenced by Origen's writings, scientific exploration of this particular miracle is not just a recent phenomenon. Johannes Kepler, the sixteenth-century discoverer of the planetary laws of motion, was the first to posit a detailed astronomical explanation for the Star of Bethlehem. In Kepler's time predicting conjunctions of planets was a major part of any astronomer's job—as important as pinning down the date of the Big Bang, discovering extra-solar planets, or finding water on Mars is today. As someone who has done his own share of celestial calculating, much of it with the aid of software packages and spreadsheets, I can scarcely imagine the time and patience Kepler needed for his mathematical machinations. With no access to calculators and computers, and

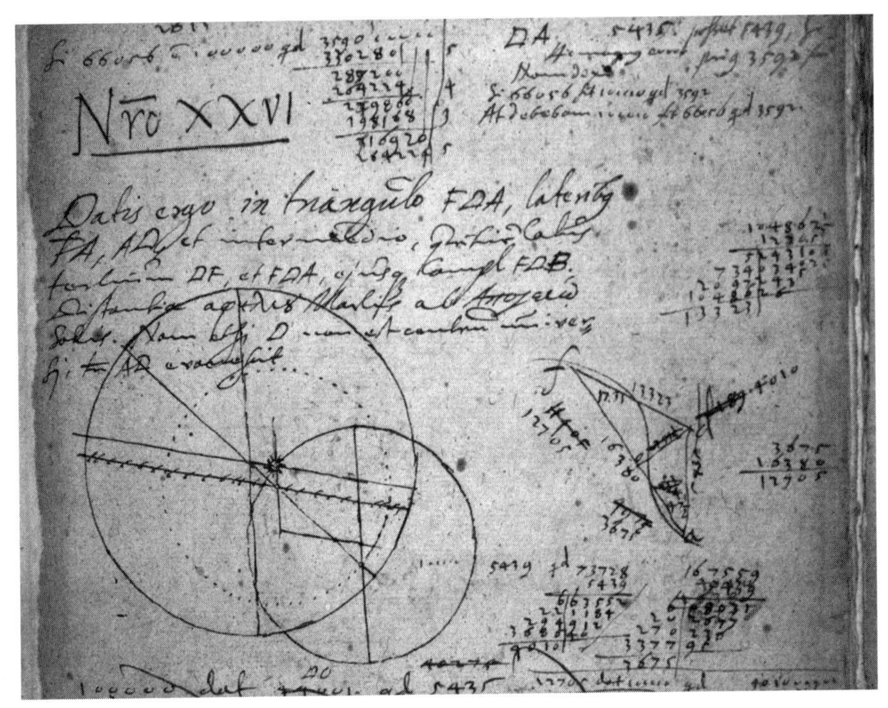

3. A page from one of Kepler's manuscripts showing his calculations. He was the first person to follow through on the conjunction hypothesis for the Star of Bethlehem. (From Arthur and Peter Beer, Kepler: Four Hundred Years, Proceedings of Conferences Held in Honour of Johannes Kepler, *268, St. Petersburg Archives of the Russian Academy of Sciences; photo by Owen Gingerich)*

deriving all the previous planetary positions from long lists in tables, Kepler had recourse only to pen and paper. I am fond of exhibiting facsimiles of Kepler's documents to my astronomy students, documents that contain monster long-division problems in the marginalia (Figure 3). Incredible! The man must have possessed a rare and deep passion and dedication to complete such a Herculean mental task.

In 1603, Kepler had been engaged in calculating the timing of a much anticipated conjunction of Mars, Jupiter, and Saturn in the constellation of Sagittarius, slated for December of that year. While checking planetary positions in October 1603, he sighted a very bright star in the constellation of Ophiuchus, the serpent holder, a mere handspan in the sky to the northwest. It turned out to be the second brightest supernova in recorded history, visible in broad daylight—even outshining the planet Venus. Taken with the almost simultaneous occurrence of these celestial spectacles, Kepler set out to write a book about the "stella nova," or "new star" as he called it, that incorporated ideas found in a book by Laurence Suslyga, a little-known Polish astronomer who had calculated that Christ had been born a year or two following a great conjunction of the same three planets in the constellation of Pisces. Such a close gathering had occurred only six times since the foundation of the world (then believed to have occurred in 4000 BC). The coincidence was literally too good to be true.

The Star of Bethlehem, if it occurred before the birth of Christ, as he wrote:

> was not of the ordinary run of comets or new stars, but by a special miracle moved in the lower layer of the atmosphere." . . . The Magi were of Chaldea, where was born astrology, of which this is a dictum: Great conjunctions of planets in cardinal points, especially in the equinoctial points of Aries and Libra, signify a universal change of affairs; and a cometary star appearing at the same time tells of the rise of a king. . . . Granted, then, that the new star of the Magi was first seen not only at the same time as Saturn and Jupiter were beheld each in the other's vicinity, namely in June of 7 BC, but also in the same part of the sky as the planets . . . what else could the Chaldeans conclude from their, and the still existing, rules of their art [of astrology], but that some event of the greatest moment was imminent? . . . Nor do I doubt but that God would have condescended to cater to the credulity of the Chaldeans.[7]

Now we know why searching out conjunctions was so important 400 years ago. Imagine how stunned Kepler must have been on that bright autumn night in 1604 when he witnessed one of the greatest supernovae of all times blazing forth in the constellation of Ophiuchus. Such a spectacular sight would have been an appropriate way to announce the grandest of all God's creations, the birth of a savior! With a touch of the unpredictable, its miraculous yet scientifically documentable aspects satisfy across the board. As one theologian has characterized it, "God becomes both grand artificer and consummate showman."[8] Incidentally, most historians seem to have lost the Kepler supernova connection, and as a result the story handed down to us suggests that the great star was a conjunction pure and simple.

Mars, Jupiter, and Saturn—a trio of planets that still captivates popular astrology. When they lined up in May 2000 doomsayers warned of a shifting of the earth's poles, earthquakes, a significant movement away from Christian principles (none of which has happened), and a dive in the stock market (this prediction did come true). That the cosmic rendezvous took place in Aquarius accorded it special significance—omens for the new age at the millennium.

Today, thanks to technology, decoding astronomical passages in the Bible can be fun—and you don't need much of an astronomy education to get involved. You simply download your astronomical software and search for unusual sky phenomena that match other Matthean phrases taken literally, like the star having been "long seen," "in the East," and that it "went before them," coming to "rest over the place where the child was."

Heading the list of modern proponents of the conjunction hypothesis is astronomer David Hughes. His scenario on a two-planet conjunction detailed in a popular 1976 book is the one adopted by most contemporary Star of Bethlehem decoders.[9] Rare Jupiter-Saturn conjunctions figured prominently in Zoroastrian millennial cosmology, which enjoyed an extensive revival during the stable period of the Roman Empire, about the first century BC. Repeated cosmic encounters were thought to represent the initiation of successive eons that made up cyclic world history.

Hughes cites a triple conjunction (three close visual passes in a row) of the planets Jupiter and Saturn in the constellation of Pisces in 7 BC, thereby placing the birth of the historical Christ around October of that year. The Magi (*magoi*), according to Herodotus a tribe of the Middle East skilled in sorcery, clearly would have been recognized by Matthew as skilled astrologers who were intimately familiar with the sky, and who already must have known what was about to transpire. A cuneiform text excavated at Sippar, a town north of Babylon known for its school of astrology, even records calculations and predictions of the event. Well aware of Jewish tradition, the Magi knew Jupiter was a lucky star and that Pisces had a strong astrological association with the Jews. Even more significant, the celestial fish are a sign of redemption and would later develop into a well-known acrostic symbol for the Savior. And if that isn't enough, the sun moves into Pisces between winter and spring, thus highlighting the contrast between the end of an old cycle and the beginning of a new one.

Hughes forcefully argues that these circumstances, taken together, would have given the wise men ample cause to make the 550-mile (three- or four-month) trek west to honor the newborn king. Moreover, the three close passages of the two planets were conveniently spread over the period from late May to early December, each reconvention confirming that the wise men were on the right track. Interestingly there are no recorded celestial phenomena that fit this period in the well-attested historical records from China.

Some modern astronomers like the one-two punch offered by combining a planetary conjunction with another supporting event. For example, the triple conjunction of Jupiter and Saturn may have focused the wise mens' eyes on the west, but perhaps it was a comet that actually commenced their journey. Comets did blaze forth in 5 BC and 12 BC. The latter, regarded by most chronologists as being a bit too early to fit biblical chronology, has been traced to one of the regularly recurring appearances of Halley's comet, historically documented since 240 BC to reappear on the average of every seventy-six years (Figure 4).

Astronomer Michael Molnar's decipherment of Matthew synthesizes astrological prediction with celestial phenomena that might not be regarded as spectacular to us (nor even to common folk in the Middle East), but which would have been important in a calculated horoscope.[10] If the wise men were seeking omens signaling the birth of a great king in Judea, they would likely have looked into what was happening in Aries, Molnar argues, for this is the zodiacal sign that ruled Judea. On April 17, 6 BC, Jupiter, the regal planet, made its heliacal rising (its first predawn appearance in the east after having been blotted out by the sun for several weeks). It happened in Aries, in the ascendant with a rising moon. Following its reappearance Jupiter also underwent retrograde motion, making a backward turn as it was joined by Mars. In terms of the contemporary astrological doctrine, all of these signs would have produced "incredible regal conditions." The clincher for Molnar's theory was a series of bronze coins—artifacts dated to AD 6–14, when the city was annexed by Antioch (in present day Turkey)—that drove his study and fueled his hypothesis. They show the Judean sign of the ram with a great star rising over its back. But Molnar's idea is even more appealing to contemporary audiences because it yields an exact date—and when it comes to decoding ancient mysteries, precision satisfies.

4. Giotto put his own spin on the Star of Bethlehem when he celebrated the 1301 return of Halley's comet, shown in the sky scene at the top. (Giotto, Star of Bethlehem, *Capilla Scrovegni, Padua, Assessorato ai Musei, Politiche Culturali e Spettacolo)*

But that hasn't always been the case. Awash in the complex timings of computer-generated cosmic loops, we tend to overlook the power of an oft-missing vital element in the star story—the role that astrology, the doctrine that our lives are influenced by the stars, might have played in the drama. More than a hint of astrology resides in Kepler's and Hughes's accounts—and especially in Molnar's. Astrology! An astronomer's blood curdles at the mere mention of the word. One planetarium director and Christmas star scrooge candidly noted that he wasn't overjoyed with several aspects of Advent star programs, "including their resorting to astrology."[11] But doing horoscopes was a way of life in New Testament times. Except in the post-Enlightenment West it has enjoyed almost universal practice. Indeed one could argue the viewpoint (as I will shortly) that resorting to astronomy may take us down the wrong path if we really want to understand the Star of Bethlehem. But first we need to explore the roots of astrology.

The doctrine "as above, so below" penetrated all levels of stratified society from noble to peasant. The perceived influence of the stars applied to all forms of activity—from politics and science to medicine and agriculture. Rome at the time of nascent Christianity was no exception, as these passages from Roman historian Livy, interweaving social events and natural phenomena, will attest:

> Between the third and fourth hour of the day, darkness set in. On the Aventine, showers of stones were atoned for by a nine-day observance. There was a successful campaign in Spain.

And:

> A nine-day observance was held because there had been a shower of stones in Picenum, and because lightning bolts, appearing in many places, had scorched the clothes of many persons by a slight blast of heat. The temple of Jupiter on the Capitol was struck by lightning. In Umbria, a hermaphrodite about twelve years old was discovered, and by order of the soothsayers was put to death. Gauls who had crossed the Alps into Italy were expelled without a battle.[12]

The last sentence in each of these passages—indeed in most astrological predictions—is the incongruous element that combats the logic of the scientifically schooled mind.

Agree with it or not, there is a logic to the underlying belief that cosmic destinies influence us and that omens issuing from the denizens of heaven affect human will. The logic behind the idea that what happens in the sky is related to what takes place here in the sublunar realm goes something like this: the ocean's tides correlate with the positions and phases of the moon—even with the sun—as any careful observer of nature will note. Many marine species breed according to lunar as well as solar cycles. And who will deny the female menstrual schedule its lunar appointment? If these celestial luminaries influence our world, might there be other tides in the affairs of humankind that bear cosmic consideration? This extension of influence to a broader range of affairs and a wider class of sky objects seems but a short extrapolation. So goes the logic—all of it embedded in a deep desire to seek order "here below."

Anyone who lives in a low-tech world contiguous to the vagaries of nature can see that what happens here on terra firma really is mirrored in the sky. Just as day turns into night and winter becomes summer, so too the rhythm of life alternates from one extreme to the other. Flowers bloom and die annually; they open and close their petals daily. Animals hibernate, awaken, reproduce, then hibernate again. Our lives oscillate between conscious wakefulness and perilous sleep, when the mind seems submerged beneath a cloud. Our moods experience good times, bad times, then good times again. And the hierarchy of movement in the sky, from plodding Saturn, perceived in the classical world by virtue of its slowness to lie at the top of the layers of heaven, to the fast moving moon at the bottom, seemed a cosmic replica of the structure of society, where kings and nobles resided at the summit of the social ladder while peasants languished at the bottom—this was *common sense.*

Given these celestial metaphors of alternation and hierarchy, the heavens became a social role model, a template that reminded us of our pitfalls as well as our opportunities. This cosmic influence does not necessarily mean, as some often assume, that sky phenomena *cause* us to behave in lustful or compassionate ways. Astrology operates on a principle of association between attributes of human and natural behavior. Such an outlook was reinforced when people realized, millennia before civilization had become agrarian and urbanized, that the movement of celestial bodies—their disappearance and reappearance, their coming together and flying apart—were, with few exceptions, periodic and pristine. And with the acquired skills of computation, they came to be predictable with ever greater precision. No wonder the Greeks thought the sky sphere was made out of quintessence, that divine fifth element of crystalline quality found only in the world above. Aristotle draws the line of cosmic perfection above the moon where, he says, the stars swing around ceaselessly, and because they are the cause of their own motion, they are deserving of the name of the "foremost and highest god, entirely immutable and owning no superior."[13] Humanity lies here below in the reflexive microcosm amidst the dregs of the universe—the cold, heavy sediment that by virtue of its weight sunk to the center of things once the gods created the universe. Our only link to the divine celestial realm resides in sympathetic forces. We must attune both our senses and our souls to them for the sake of our own betterment.

The Chaldeans, ancestors of the wise men, thought cosmic influence derived from the complex interaction of celestial spirits in a living sky. We mortals here below, being an interactive part of the cosmos, vibrate sympathetically in response to their heavenly emanations. The emanations can strengthen or weaken according to where the wandering lights, the planets that represent the gods, travel in the sky and—this is especially significant in the Chaldean system—at what time of year and time of night they rise or set on the local horizon.

The Chaldean astrological system allocated different parts of the earth to particular regions of the sky: north to Akkad or Babylonia, south to Elam, west to Syria and Palestine, and east to Assyria. Where a planet appeared, how long it stayed its course, where it made its cyclic retrograde or backward turn—all of these indications—offered a variety of interpretations for what I like to call the "When . . . then" statements written in cuneiform by the attentive astrologer; for example, "on the twenty-fifth day of the ninth month [Venus] disappears in the east . . . then the harvest of the land will be successful."[14]

Infrequent, and therefore harder to predict, close encounters of the planets (such as when Mars crossed paths with Jupiter or when Venus confronted Saturn) were especially influential, often portending extraordinary nativity, as we shall see. A nova, or "new star," suddenly flaring up to daylight visibility for a few weeks also offered an obvious celestial metaphor signifying something different and unusual from the omen-bearing luminaries. A rarely observed solar eclipse, when the moon covers the sun and darkness reigns in the daytime, might hold multiple prognostications. An ancient Chaldean omen reads: "When an eclipse begins on the first side and stands on the second side, there will be a slaughter in Elam." But such predictions were not necessarily always bad: "When an eclipse happens in the morning watch . . . [with] a north wind blowing, the sick in Akkad will recover." Once enough such events were observed to discern a pattern, then the astrologer gained power, thus: "On the 14th an eclipse will take place; it is evil for Elam and Amurra, lucky for the king."[15] But once again we must remind ourselves that these happenings were not thought to be the effect of an inalterable set of scientific laws of cause and effect, as our common sense, in hindsight, would dictate; rather they were considered voluntary acts in an animate cosmos committed with the same sort of intent and backed by the same sort of intellect as that of any mortal down here on earth.

These articles of faith in foretelling the future by inspecting events in the cosmic domain, which paralleled all terrestrial affairs, penetrated from the east into Hellenistic astronomy after the age of Alexander, which saw a rapid mixing of diverse cultures and ideologies. In the third century BC, the Babylonian priest Berossus taught the history of his culture on the island of Cos, and a century later books on astrology by Babylonian authors were read by the Stoic school. In the Greco-Roman world the cosmic sympathies were applied to the affairs of individual commoners as much as to those of kings.

We owe our word horoscope (from *horoscopus* = "I observe the hour"; literally, "I watch what rises") to the democratically oriented Greeks. Basically everyone had a right to know the future. The nativity horoscope—which gave birth to modern astrology, or the art of predicting general patterns preprogrammed into your future life—is based on an examination of celestial bodies appearing over the eastern horizon at the time and place you were born. The more social reality denied the cosmic appearances, the greater the effort on the part of the astrologer to re-envision and revise the rules. And the greater the astrologer felt the need to "save the phenomena," as the Greek philosophers put it—that is, to generate models capable of yielding even more precise calculations to pinpoint future astral events.

At the pivot of public interest, astrology was, as historian of astronomy Anton Pannekoek has described it, a paramount "theoretical doctrine of scientists and philosophers" by Roman times.[16] Works on astrology at the beginning of the Christian era include Manilius's Latin poem on the *Circle Theory of the Celestial Phenomena.* Almanacs filled with planetary positions, and "When . . . then" predictions in Greco-Babylonian style ran rampant in an empire open to influence from the cultural periphery. A celestial globe was among many artifacts brought to Rome from Sicily by the consul Marcellus. The Farnese "Atlas" (first century AD) depicts only the figures and not the component stars of the constellations, probably because astrologically important effects were thought to emanate from entire constellations rather than individual stars. The whole of the sky globe rests on the shoulders of the heroic world-bearer.

Early Christians initially debunked astrology. St. Augustine's fourth-century AD *Confessions* was rife with stories alluding to astrologers' illusory claims and insane rituals. Opposition arose not because there was anything inherently wrong with astrology but simply because the Greek philosophers who practiced the art were considered pagan. If astrology was part of heathen wisdom, then it must oppose the will of God. But there were aspects of astrological doctrine that appealed to many early Christian sects; for example, the doctrine of truth by revelation and the dialog that ensued between priest and client. This led to compromises; for example, in much later medieval astrology the stars were consigned to rule the body while God guided the soul.

The Christians, once established, worked to appropriate those aspects of astrology along with the attending celestial imagery that supported their doctrine. As believers were weaned away from pagan polytheism, the planets were transformed into mediators who announced the intent of God but were never allowed to act contrary to his will. Altered by the new garb of Christian morality, the pagan sky gods survived. Ecclesiasts altered the characteristics of old Babylonian Mercury, a speedy seeker of earthly wisdom, making him over into a pious and scholarly scribe. Jupiter, once a judge, is portrayed in Christian art as a monk with a chalice in one hand and a cross in the other. Perhaps most recognized of all the old wanderers, Venus, goddess of love, became the image of Christian love portrayed in great works of art well into the Renaissance.

Religious syncretism carried over into the celebration of festal dates. In AD 273 the emperor Aurelian officially designated December 25 the Roman winter solstice (literally, "solar standstill") festival. This day was already the Birthday of the Unconquered Sun (*Dies Natalis Solis Invicti*), so called because of what actually happened in the sky on that occasion. This feast signaled the time when the sun god Mithras/Apollo would not vanish on his southward winter course but instead would remain in the sky and achieve victory over darkness. Here was time's visible pivot, a vital, easily discernible turning point between alternatives. The sun starts back on its northward course toward humanity, bringing an end to the menace of the dwindling daylight hours and consequently greater warmth and longer light. Winter solstice was a day of joy and hope heralding the glory of springtime. Believers realized that the light from heaven, although somewhat diminished, is nonetheless eternal. When the fourth-century emperor Constantine made Christianity the official religion of the Eastern (Holy) Roman Empire, it did not take long for this feast to be adopted as the celebration of the *Dies Natalis*—the day the light of Christ comes into the world to restore life and hope. Today the lighting of the Christmas tree, with the Star of Bethlehem on top, and the burning of the Yule log (from *hiaul* or *huul,* Old English words for "wheel" or "sun") are remnants of rituals in which celebrants beckon the holy light to return.

Now that we know something about astrology and the central role of the sky in the lives of everyday people in classical antiquity, we can better appreciate the powerful symbolism once borne by celestial light—and this gives us a better perspective on what the Star of Bethlehem's story might have meant in its time, when great earthly events required astrological portents. But despite having gone this far, non-astronomical perspectives on the star story are still lacking.

Because science rules our day, the main focus in the study of the star at the top of the Christmas tree seems to center around defining and describing it. There are, however, other points of view, other questions. For example, if the astronomer's question about the great Christmas event is, *what* was that star? the historian's question might be, *was* there a star? Could it have been part of a myth? We tend to think of myth as veiled or fabricated truth, a product of the fanciful imagination—something just waiting to be debunked by science and replaced with literal truth. But for people who tell their myths to relate historical events to beliefs and practices in their religious systems, there is truth in myth. For example, timeless festivals like the Birthday of the Unconquered Sun have been celebrated around the world to rejoice that the sun has reached its turning point in the winter sky. Although I would not consider myself a religious person, I radiate joy when, after a long, hard winter, I begin to see the days warm up, the snow melt, and the first green plants sprout in our front yard with the increasing light that comes back to my environment once the sun passes winter solstice. Truly this is my New Year. The Egyptians marked theirs with the heliacal rising of Sirius; the Aztecs with the passage of the sun overhead; the Romans with the *Dies Natalis Solis Invicti,* and the early Christians took it from there.

Modern scientists might choose to tell the story of the great solar turnaround graphically and mathematically. They might present charts showing the variation in the number of hours of daylight through the winter months; they might calculate the insolation (the quantity of solar energy incident on a unit of landscape at various latitudes at different times on different dates). They could correlate these data with the times of germination of various plants, the end period of hibernation of bears and beavers, daily temperature maxima and minima, rainfall records, and so forth. Thus emerges a rationalist version of the story of the turning of the solar orb, and it contains many truths. But there are many ways to reveal truth. The turnaround can also be told mythically, perhaps by personifying wind and rain, sun and fertility, maybe even by garbing these attributes of nature in the clothing of gods and goddesses. The language of myth is different from the language of science. Its narrative is poetry and imagery; its grammar, analogy and metaphor.

Quite contrary to scientific storytellers, the best mythical storytellers are always careful to craft combinations of words and descriptions designed to give the listener or reader the strongest indication of what the great turning point *feels* like—all the better to move us to celebrate it. My wind god would have puffy cheeks and a large chest and my fertility goddess would be voluptuous. I'd make the celebration a communal potluck, to which each celebrant brings a dish appropriate to the season. I'd bring the first fruits and veggies of spring (like asparagus and peas) and of course new wine—a crisp Beaujolais would be ideal. And I'd accompany the meal with a series of toasts, brief words describing what the big turnaround means to them in the great cycle of life to be uttered by all participants—and, yes, maybe a song or two. (I love Beethoven's Ninth, particularly the last movement.)

Whether the story be literal or metaphoric, scarcely a culture in the world does not recognize and celebrate the advent of the returning light. But would you say that my "pagan" account of the great solar turnabout holds no truth? Don't its symbols—what we eat, what we say, sing, or do, convey the idea and reality of the renewed light that dawns upon us when the sun changes it course? The dispassionate scientific account of the winter solstice event may be more precise and detailed, but the mythic version is far easier to relate to—at least for most of us. Myths certainly aren't "all made up" as uninformed critics say. The Star of Bethlehem is an Advent myth, and that we celebrate it at solar turnaround time is particularly significant.

If we think of Matthew's story as a myth, naming the precise celestial event that set the Magi on their journey isn't really so important, is it? The efficacy of the story lies in getting across the idea that there was a group of disenfranchised people who lived 2000 years ago. On a night of glorious hope they were desperately seeking signs of a better future that would surely come with the birth of new light upon the world. Finding the star would have had little to do with connecting the believer to the significance of Christ's coming.

So, was there a star? Maybe there was, maybe there wasn't. My own opinion is that there probably was some sort of sky event. It may not have been dazzling to the eye (as Molnar has suggested), and it may even have happened after the fact. (Did you know there was an eclipse of the sun the day after the death of Princess Diana?) Verifying the absolute truth about a sky phenomenon portending the Advent may be about as important as seeking an optical, electromagnetic explanation for the radiated light in Christ's transfiguration—or finding the exact spot where Edvard Munch stood when he painted "The Scream." The astronomer and the historian hold different points of view. They ask different questions because the knowledge each holds dear is not the same.

"Suddenly I had an epiphany!" a friend told me when he finally thought he had grasped the meaning behind an ancient Mexican ritual he had been studying. But an epiphany is an appearance or a revelatory manifestation of something divine, a supernatural event—a miracle. He may have had a "eureka moment," but he certainly didn't have an epiphany in the true sense of the word.

Do you believe in miracles? There isn't much room for miracles in today's world, but the case for a supernatural event beyond all scientific analysis—once the standard explanation for the Star of Bethlehem—remains plausible, at least to some. Can we second-guess the Creator? Why should a believer even be tempted to look for a scientific explanation? Had God been so disposed, he could have created a heavenly baby for any purpose. But the ruler of the universe is frugal, goes the rational counterargument: surely he would rather have made use of cosmic arrangements he had already fixed in the heavens to deliver his message. Penetrating the mind of God is no mean task. No wonder modern science finds the miracle explanation theologically weak. Whether they believe in a deity who set natural laws in place or in no deity at all, scientists are conditioned to think that a perfectly natural phenomenon must have occurred that can account for all the information given in the historical sources—that's pure twenty-first-century common sense.

By contrast, humanists stress the mythic aspects and theologians the role of miracles in the scriptures. "[A] Miracle is simply what happens in so far as it meets people who are capable of receiving it, or are prepared to receive it, as a miracle," wrote theologian Martin Buber.[17] When we try to dismantle an omen in search of its underlying causes, Buber argues, we can often lose sight of the meaning it was intended to convey to the true believer who experienced the sign. Although some have attempted to reconstruct the natural events that gave rise to the story of Moses's parting of the Red Sea, how the tides work in the Gulf of Aqaba is irrelevant to the far more important question of how the children of Israel interpreted whatever happened. Here the theologians seem to be asking an even more basic question than the ones asked by the astronomers and historians. They want to know, what does the story of the star in the gospel mean? What's important to them is that there *is* a story.

For those who follow the Christian way, the "star event" becomes an abiding pillar in the edifice of their coming together as a people. I find it interesting that scientists react negatively to the word *story.* Isaac Asimov once pronounced the Star of Bethlehem "*merely* a wonder tale" (my italics).[18] And Carl Sagan tells us flat out that "Matthew botched the story. He was too vague. All he said was that the star appeared in the east. He didn't even tell us which constellation or anything."[19] These critics from the hallowed halls of logical positivism are not concerned with the meaning of the story. But if the story related by Matthew is "just a story," that does not mean that it is devoid of truth.

Theologian Kim Paffenroth explains the famous reference in the gospel of Matthew as *Midrash,* a method for arranging truth through story as old as the Talmud. In his view the miracle star narrative is a tale that reveals what whoever wrote Matthew's gospel felt to be the truth about a man taken to be Christ. The writer, following the style of the times, was concerned with neither historical literalism nor exact science—at least not as concerned as many of us seem to be. The narrative of Christ' s infancy is really a story about the good news of salvation, and only that. The story is to be appreciated only by those who have the eyes to see it. Therefore, we should not be concerned with reading Matthew's gospel in any other way, lest we do violence to his account.

Historians of the New Testament Bible find a stark parallel, some call it a prophecy, to the star story in the Old Testament. When Moses was leading the Israelites to the promised land, he encountered a magus who told him: "There shall come a man out of Israel's seed, and he shall rule many nations. I see him, but not now . . . ; a star shall rise from Jacob, and a man shall come forth from Israel."[20] Theologians interpret that star to be David and the myth to foretell the reign of his dynasty over the kingdoms of Israel and Judah. Later this passage was reinterpreted to imply that there would appear a new Messiah descended from that dynasty. Gospel means glad tidings, good story; and that's just what it is, a story designed to convey why believers should view Christ as the great salvation of the world. Paffenroth, who has chronicled a history of viewpoints about the Star of Bethlehem, notes that, in stark contrast to the literalist twentieth- and twenty-first-century accounts, those of the nineteenth century tended to dismiss the star as pure fiction. So maybe the question shouldn't be, what really happened? Rather, we should ask, what does it mean for Jesus to be the savior? To be the Christ, as to have been any shaker or mover in those times, means to have been born miraculously under a guiding star. It takes a revelation to make a Christ.

From the point of view of the nonbeliever, a miracle—an epiphenomenal event—is offensive. For the scientifically minded, it is a fabrication of pure imagination; from the historian's viewpoint, it is irrelevant; but for the believer—the one who possesses the faithful eyes to see it—it is the ultimate salvation. Who is correct depends on who wants to know.

Some ideological conflicts between faith and reason, such as the one concerning the nature of the unborn fetus, are probably worth arguing over. After all the question of when life begins is rife with social, political, and legal issues that directly affect our lives. But who wants to engage in an ideological dispute over a star? Yet the conflict is the same, even if the debate is at least superficially more measured in areas where the stakes are lower .

And a conflict it shall remain, for when it comes to science's take on the natural world there are basically two points of view. The reductionists, on the one hand, harbor the belief that true knowledge of the world—regardless of the discipline that acquires it, be it physics, chemistry, biology, archaeology, psychology, anthropology, or even philosophy or religion—ultimately can be reduced to science. Whatever can not be so distilled is simply not worth knowing. Diametrically opposed to the reductionists are the traditionalists. They believe that knowledge emanates from many sources and that science is but one of them. For a traditionalist, ethical, artistic, moral, and religious knowledge cannot be judged to be of greater or lesser value than scientific knowledge. As author-physiologist Jared Diamond puts it, they "stand side by side with science, parts of a human heritage that is older than science and perhaps more enduring."[21]

I have not conducted a survey of where people situate themselves along the R-T (Reductionist-Traditionalist) spectrum of beliefs (if I may call it that); but to judge by the way they address the Star of Bethlehem question, I would argue that there is a deep divide. A fair percentage of working scientists lean far R, but the remaining seekers of knowledge are on the far T side of the spectrum. I suppose a "moderate" might argue that although other forms of knowledge are worth knowing, in the end scientific knowledge is all that we can really depend on.

We live in a time when the drive to reduce all forms of knowledge to science thrives—from the decline of talk therapy in favor of drug therapy to the proliferation of genetic and chemically based explanations for most forms of human behavior. Somewhere in the midst of the great divide between these two beliefs lies the quest for true knowledge about the Star of Bethlehem. Whatever point of view we acquire, it will always be conditioned by our own prejudices, colored by an ignorance of the depth and complexity of the beliefs of other people who lived in different times. Sooner or later most of us achieve that level of comfort we all seek when we settle on what we believe. And so we reach for the stars. But once we try to touch the Star of Bethlehem, we find that, like the rainbow's end, it vanishes before our eyes. Like searching for unicorns, the quest for the Star of the Bethlehem tells us more about what lies in ourselves than in our stars.

NOTES

1. Origen, AD 248, *Contra Celsum* 1.58, in *Origines Werke,* ed. Paul Koetschau (Leipzig: J. C. Hinrichs, 1899), 1:109.

2. R. J. Russell, Center for Theology and the Natural Sciences, Berkeley, reported in A. J. Orion, ABCNews.com (March 18, 2004).

3. Bruce Feiler, *Walking the Bible: A Journey by Land Through the Five Books of Moses* (New York: Morrow, 2001).

4. Dan Brown, *The Da Vinci Code* (New York: Doubleday, 2003).

5. Matthew 2:1–2.

6. Colin Humphries, "The Star of Bethlehem—a Comet in 5 BC—and the Date of the Birth of Christ," *Quarterly Journal of the Royal Astronomical Society* 32 (1991): 381.

7. Christian Frisch, ed., *Joannis Kepleri Astronomi Opera Omnia* IV (Frankfurt: Heider & Zimmer, 1858), 346–347. Cf. W. Burke-Gaffney, "Kepler and the Star of Bethlehem," *Quarterly Journal of the Royal Astronomical Society* 31 (1937): 417–425.

8. Kim Paffenroth, "The Star of Bethlehem Casts Light on Its Modern Interpreters," *Quarterly Journal of the Royal Astronomical Society* 34(1993): 454.

9. David Hughes, *The Star of Bethlehem: An Astronomers' Confirmation* (New York: Walker, 1979).

10. Michael Molnar, *The Star of Bethlehem: The Legacy of the Magi* (New Brunswick, NJ: Rutgers University Press, 1999).

11. Quoted in Robert Berman, "Night Watchman," *Discover* 11 (December 1990): 78.

12. Arthur Schlesinger, ed. and trans., *Livy* 14 (Cambridge: Harvard University Press, 1959), 248.

13. Aristotle, *Physics,* ed. and tr. Philip Wicksteed and Francis Cornford (London: Heinemann, 1939), i:9.

14. Peter Huber, "Early Cuneiform Evidence for the Existence of Venus," in *Scientists Confront Velikovsky,* ed. Donald Goldsmith (Ithaca: Cornell University Press, 1977), 123.

15. R. Thompson, *Reports of Magicians and Astrologers of Ninevéh and Babylon in the British Museum* 2: iii (London: Lusac, 1900), 271.

16. Anton Pannekoek, *A History of Astronomy* (Toronto: Allen and Unwin, 1961), 131.

17. Martin Buber, "The Wonder on the Sea," in Martin Buber, *Moses: The Revelation and the Covenant* (New York: Harper, 1958), 74–79.

18. Quoted in Peter Brown "Star of Wonder," *Connecticut* 38 (1990): 46.

19. Ibid.

20. Numbers 24:7, 17.

21. Jared Diamond, "Twilight at Easter," *New York Review of Books* (March 25, 2004), 5; whence the term *traditionalist.*

TWO

WHERE ORBITS CAME FROM AND HOW THE GREEKS UNSTACKED THE DECK

Let no one enter here who knows no geometry.

ALLEGED INSCRIPTION ABOVE THE DOOR OF PLATO'S ACADEMY[1]

[Hippodamus was] a living example that illustrates the connections between one man's astronomical preoccupations concerning the celestial sphere, his search for the best political institutions, and his attempt to build a town according to a rational, geometrical model.

–CLASSICIST J.-P. VERNANT[2]

In a scene from the film *My Big Fat Greek Wedding*—a comedy about the loss of ethnic pride and the difficulties experienced by immigrants struggling to adapt to urban life in a Chicago suburb—Gus Portokalus, the beleaguered patriarch of a Greek American family, instructs his preteen daughter, Toula, and her friend while taking them to school. His topic: the many gifts bestowed upon the world by ancient Greek culture, especially the words we speak. "Give me a word—any word," he says, "and I will show you how it came from the Greeks." After his passengers fail to respond following a long silence, Gus offers *arachnophobia*. "It comes from *phobia*, 'fear,' and *arachne*, 'spider.' So, you see, 'fear of spiders'; it is Greek!" Other examples follow, but Toula's and her friend's rolling, adolescent eyes tell the viewer (although perhaps not Gus) that the lesson is wasted. Establishing one's ancient heritage evidently does not loom high on the list of priorities that engage the minds of the young.

But Gus is correct, even if he is a bit comical: ancient Greek culture, like no other, lives on within us. We rediscover them in our tragedy and comedy, in our aesthetic comprehension of art, in our concept of the ideal, in our system of formal logic, in our mechanical view of the universe—even in our college curriculum. I've always found irony in our use of the word *trivia*—from the Greek word for crossroads—to imply something common or dispensable when, in fact, the word once defined the core of studies (grammar, logic, and rhetoric) in the medieval university.

We also find in ancient Greece the roots of our contemporary common sense about how we picture the extraterrestrial environment. *Planet,* for example, is another Greek-rooted word. It comes from the noun *planétés,* "wanderer," and the verb *planein,* "to wander." That's just what a planet does as it pursues its path among the stars: it wanders among the constellations that make up the zodiacal highway. Who can forget the order of the planets that wheel around the fiery luminary at the center of the solar system? Mercury, Venus, Earth, Mars, Jupiter, Saturn, Uranus, Neptune, Pluto—I learned that in the fourth grade. I can still remember making clay models of each and placing them on penciled-in orbs on a giant sheet of corrugated cardboard fashioned out of a dismantled carton. Because Jupiter is thirty times the size of Mercury, I always had a problem getting enough raw material to roll giant Jupiter- and Saturn-size balls without reducing the Earth to a sphere of sub-peppercorn dimensions—and Mercury to a mere speck. Of course, I would have required 27,000 times more clay by volume to mould Jupiter to scale. Getting the spacing right was another problem, given that Jupiter's orbit needs to be 4,000 times that planet's diameter from Earth's orbit to represent reality. That's 2½ football fields for a golf ball–size Jupiter! Needless to say, my cardboard models were not very precise.

We cannot imagine any other way to conceive of the solar system than in terms of orbiting spheres in a vast three-dimensional sea of space. That's what the Greeks handed down to us. On a lesser scale we think of satellites orbiting planets, and on a larger scale our sun, along with 200 billion other stars, orbits

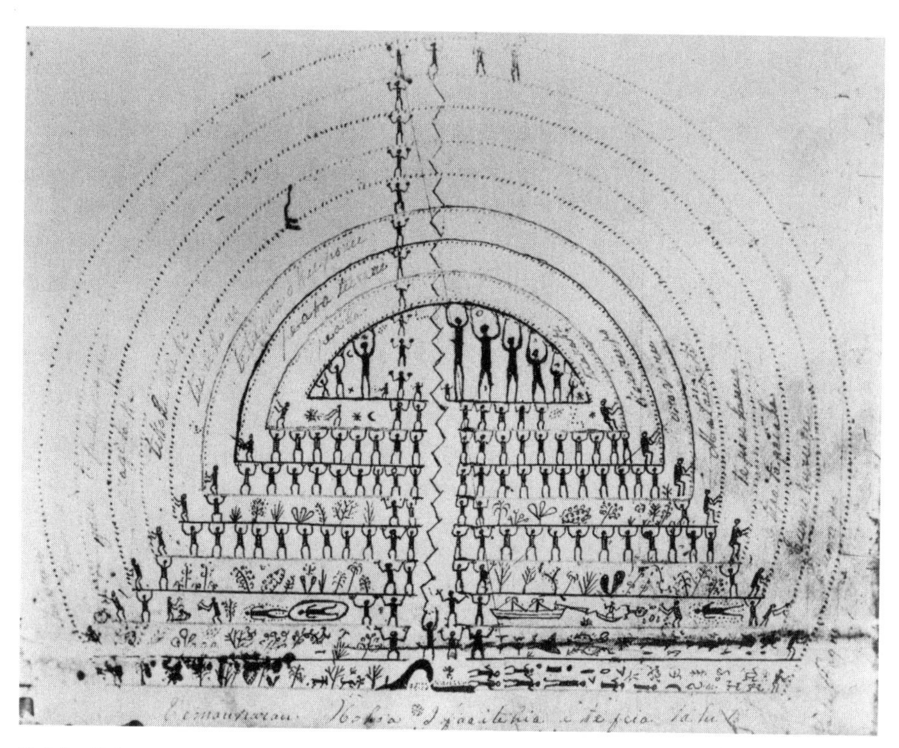

5. The Polynesian (Tuamotuan) universe. Note the channels of communication where land and sea meet sky. (T. Henry Ball, Bernice P. Bishop Museum, no. 48, Honolulu)

the center of the Milky Way galaxy some 25,000 light-years distant. Small things wheeling about bigger things in their orbits: that's the real truth about the universe. It's common sense, and anyone who imagines it to be different is delusional. But there are other possible ways and they are worth thinking about, not because we dare deny our cherished gift from the Greeks but because we want to get a perspective on why they packaged reality for us this way.

For example, like most Polynesian models (Figure 5) the ancient Hawaiian universe consisted of several concentric sky domes (*kahiki*). These domes were stacked one upon the other and centered on the Hawaiian islands, which were regarded as the center of the universe. Openings at the horizon served as channels of communication. *Kahiki-moe* was the zone of the earth that stretched as far as the eye could see. It included all the islands that lay within its boundary. *Kahiki-ku* was the portion of that zone that bends upward from the horizon and it contained the clouds. Other *kahiki* encompassed more distant, less important islands and sky objects. We can think of them as spheres of influence.

The Aztec universe was layered too. It positioned Ometeotl, a half male, half female creator deity, in the highest layer of heaven and relegated the lesser powers to lower tiers of the cosmic house (Figure 6). The Maya universe had thirteen layers of heaven surmounting nine layers of the underworld, each ruled by a Lord of the Night. The earthly plane was sandwiched in between heaven and the underworld.

Of all the cultures that thrived on the two American continents before Columbus's arrival, the Maya without a doubt came closest to rivaling the cultures of the ancient Middle East when it came to making mathematically precise predictions about what would transpire in the heavens. Epigraphers have already deciphered a table in one of the Maya's few surviving bark-paper documents. It dates from the fourteenth century and it follows the movement of the planet Venus to an accuracy of one day in 500 years. An adjacent eclipse table gives warnings of impending lunar and solar eclipses. Still more recent decodings of these texts, all written in Maya dot and bar (Base-20 notation), reveal an intimate knowledge of Mars. Maya astronomers seemed as interested as their Renaissance contemporaries in following the intricate retrograde or periodic backward motion of the red planet against the background of zodiacal constellations. As you might guess from reading Chapter 1, their principal motive was astrology. But their stacked model of the universe had little to do with the task of merely making predictions, which were all done by arithmetic. Rather, the layered Maya universe served as a colorful backdrop for telling the story of creation. In a series of episodes, hero twins and their descendants make repeated descents into the underworld to do battle with the gods of pestilence. Having conquered the terrible ills that would otherwise plague us, the twins finally rise into the upper layers of heaven where they become the sun and the moon. The layered universe is a theatrical backdrop—a vehicle—to articulate desirable qualities inbred in the first humans.

The Navajo were little different. Diné Bahané, their creation story, is a myth of emergence. Various forms of life pass upward through holes in a series of time-space spheres in an evolutionary kind of hierarchy, transforming themselves into ever more civilized forms as they ascend. Thus, the insects who dwell in the lowest order must fly upward to become air-spirit people. In the

6. A Central Mexican universe. Each layer harbors its own cosmic denizens, and a male-female creator resides at the top. (Codex Vaticanus A., 3738, 1v, Akademische Druck-und Verlagsanstalt, Graz Austria)

process, they become swallow people, then grasshopper people, then primitives, and so on, each residing in an ever higher sphere in the hierarchy. Emerging through a hole in the sixteenth sphere, fully domesticated First Man and First Woman safely arrive on the surface of the Fifth World, our present existence in time.

As in the Maya universe, the underlying action-based theme about the transformations that unfold in the stacked Navajo space-time spheres is morally based. The storyteller portrays each re-creation, each movement from layer to layer, as a correction that helps bring about order and harmony. For example, the swallow people learn to accept authority and they begin to treat one another as members of a tribe—a behavioral trait not possessed by their predecessors. When first confronted by newcomers, they react not with hostility and suspicion but with reasoned trust and acceptance:

> "Until you arrived here, no one besides us has ever lived in this world. We are the only ones living here."

The newcomers respond by pointing out that they are like the swallows in many ways:

> "You understand our language."
>
> "Like us you have legs; like us you have bodies; like us you have wings; like us you have heads."
>
> "Why can't we become friends?"

The swallows respond:

> "Let it be as you say," they replied.
>
> "You are welcome here among us."

The storyteller continues:

> So it was that both sets of people began to treat each other as members of one tribe. They mingled one among the other and called each other by the familiar names. They called each other grandparent and grandchild, brother and sister; they called each other father and son, mother and daughter.[3]

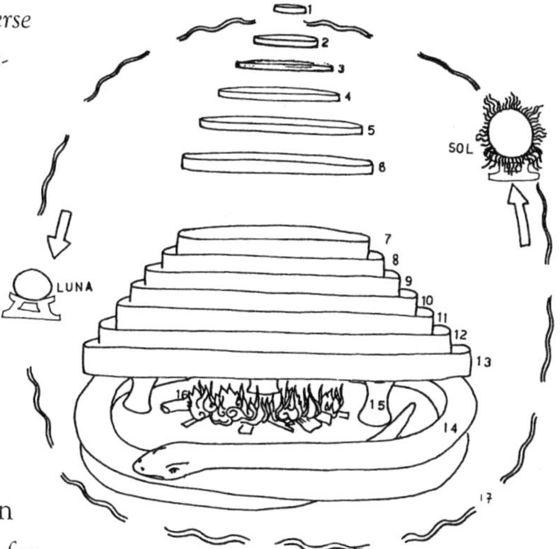

7. The Amazonian version of the universe features a coiled-up anaconda protecting a perpetual fire beneath a thirteen-layer world. We live in the seventh layer. (Jorge Arias and Elizabeth Reichel, eds., Etnoastronomias Americanas *[Centro Editorial Universidad Nacional de Colombia, Universidad Nacional de Colombia])*

These exotic universes I've been describing (see Figures 7 and 8 for more examples) have a lot in common. Their layered design was motivated in large measure by the lived experiences of the societies that conceived them. Such realities seem a far cry from the universe of Voyager, Mars Explorer, and the International Space Station. But the same is true of the original world view handed down to the ancient Greeks by the Babylonians.

Try to imagine an old Middle Eastern family patriarch—much like Gus in My *Big Fat Greek Wedding*—adapting to Greek life in fifth-century BC Athens teaching his kids: "You give me one idea and I will show you how it is Mesopotamian." What we owe to the Greeks, we also owe, at least in part, to the Babylonians, especially in the science of cosmology. They were the inventors of arithmetic, at least on the far side of the Atlantic. It was the combination of the Babylonian quantitative, arithmetical approach to planetary motion, together with the Greek logical, geometrical way of thinking that gave rise to our present understanding of the universe.

The Babylonians had their own ideas about how to model the universe. It too was stacked—a series of concentric spheres arranged one upon the other (Figure 9). Each sphere contained its planetary inhabitant and each rotated about a fixed earth in a particular period. There was nothing abstract about this model. The planets were living cosmic powers, each with its own particular influence. Marduk was Jupiter, king of the Gods; Inanna (Ishtar), Venus, the goddess of love; and Sin, the moon god.

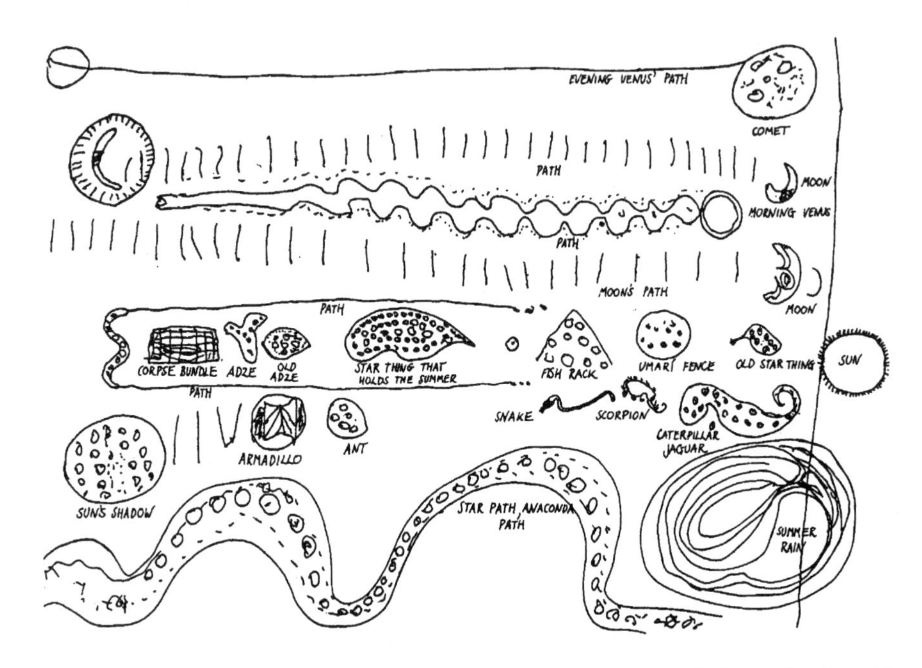

8. *The contemporary Barasana (northwest Amazon) universe consists of a layer of paths, with each component of the universe belonging to a particular layer. (Stephen Hugh Jones, "The Pleiades and Scorpius in Barasana Cosmology,"* Annals of the New York Academy of Science *385 [1982]: 187)*

Where did they ever get that idea? *As above, so below.* The Mesopotamian cosmos was a mirror image of society and it was run just like the city-states, according to a hierarchy of power. The gods were positioned at the top of the heap, and lowly humans, the equivalent of slaves, lay at the very bottom. (Most textbooks point out, quite erroneously, that ancient people of the Middle East must have been highly egocentric because they placed the earth at the center of everything.) Quite simply, the Babylonian system of astrology based on the order of the spheres shows how the design of the cosmos imitates the politics of the city-state. For example, Saturn was the planet that occupied the highest sphere, the "sphere of greatest magnitude," as they put it. It, therefore, wielded the greatest power or influence over humanity. Why should Saturn occupy the highest sphere? Just look at how long it takes to change its position among the constellations of the zodiac—longer than any other planet. Slightly faster moving Jupiter was next down the ladder of influence; then came Mars, the

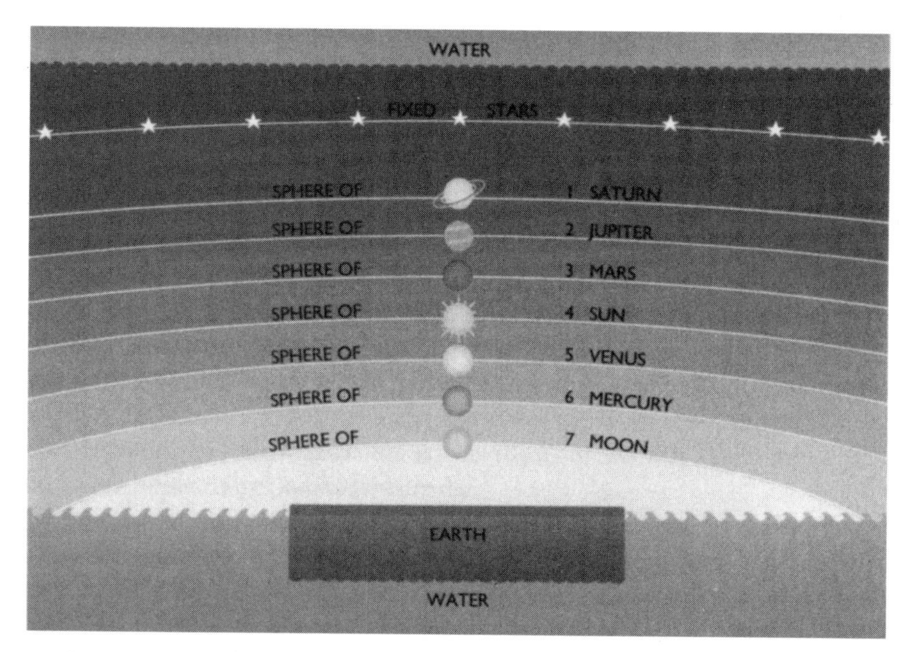

9. The Babylonian universe placed the earth at the center of a water-bound hierarchy consisting of stacked layers; each planet occupied its own layer. (From Anthony Aveni, Ancient Astronomers *[New York: Harper Collins, formerly Smithsonian, 1993], 13)*

Sun, Venus, Mercury, and, in the lowest sphere, the moon—fastest of all. According to this model the power of celestial influence waned as one moved down the hierarchy from top to bottom. So too, as we learned in Chapter 1, the nature of the power inherent in each planet was believed to alternate. Thus, Saturn was considered the most powerful evil, Jupiter the most powerful good, Mars evil but less potent, and so forth. But isn't that just like life? Darkness follows light, good days follow bad days, the seasons wax and wane from wet to dry, from cold to hot. There is no apartheid in the Babylonians' cosmos. Their universe was a reflection of life. They patterned it after their society.

So did the early Greeks. Homer's *Iliad,* written in the eighth century BC, gives a description of a shield forged for Achilles by Hephaestus, god of fire. The front of the shield depicts a map of the cosmos. The earth, sea, sun, moon, and stars are positioned at the center, and the great river-ocean circumscribes the rim. In between are scenes of human activity: the city at peace and at war, plowing and reaping, sheep raising and ceremonial dancing. Representing human action in a broader, universal context appears to be the idea. In the cosmology of Homer, the physical and human realms are inseparable.

With the exception of fourth-century BC Aristarchus, the Greek philosophers put the earth (rather than the sun) at the center of the universe. Modern science has made a big deal out of this particular error of their ways. It was not until the late fifteenth century that Renaissance astronomers replaced earth with sun at the universe's center, drastically transforming the Greek model. A point not to be lost here is that the Greeks were already thinking orbits; they were the first to shift the emphasis from things residing in layers to things in motion about a center. They unstacked the deck.

The Greeks were different from other ancient cultures in the way they thought about the heavens. Western science is a product of what historian of science Derek Price calls the "Greek miracle," a unique way of understanding the natural world based on the dominance of number (see Chapter 5), derived from the Babylonians and transferred, via logic, to geometrical imagery. This innovation was accompanied by an intense focus on motion in the heavens and the consequent development of an abiding faith that a rational basis could be devised to explain it.

Why these circumstances took place in the circum-Aegean area in the first few centuries BC is still widely debated. The Greek democratic attitude certainly helped condition them to explore the exotic mathematics that came from the Middle East, to openly debate different theories of motion, and to tinker with models that would precisely duplicate the orbits they believed gave rise to the motion they perceived in the sky. But chance also played a pivotal role in the coming together of so many of the approaches and skills that produced a major part of our scientific legacy. This peculiarity, this fundamental difference between the forebears of our civilization and all others, argues Price, was "surely a spectacular accident of history."[4]

Orb means circle, and it was Plato who first proposed that all heavenly bodies move in circles at an unchanging rate. He handed his mathematically endowed followers the following problem (as phrased by the sixth-century AD Platonic commentator Simplicius):

> What circular motions, uniform and perfectly regular, are to be admitted as hypotheses so that it might be possible to save the appearances presented by the planets?[5]

For the Greeks, "saving the appearances" meant predicting the location of a planet in the future based on records of its past motion, and doing so as precisely as possible. To the victor go the spoils. Whatever combination of geometric constructions of circles it comprised, the best hypothesis was the one that yielded an orbit that put a given planet precisely on schedule as seen by the eye. Astrology demanded this precision.

The model of Eudoxus was among the early contenders. He imagined a series of hard, nested spheres each with its own planet appended to it. (What else could hold a planet in the sky?) A huge crystalline sphere that carried the stars encompassed all of them. The earth lay at the center. Yes, this model is a series of stacked spheres, but the emphasis lies less in the stacking—the static skeletal structure that holds things together—and more in the dynamics of the planets that move with the spheres. The motion is circular. And it can be adjusted by changing the angles between the axes of the spheres and the rates at which they turn. Eudoxus required twenty-seven moving spheres to achieve his goal of saving the appearances.

In search of an extra credit project to redeem a failing grade in my beginner's astronomy course, one of my more mechanically inclined students persuaded me to allow him to construct out of wood and metal a table model of the Eudoxian spheres. After several days struggling with hoops, plates, and gears, he realized the complexity of the task, gave up working with his hands, and went back to the books, but not before he learned firsthand why the Eudoxian model was doomed to fail. He passed the course.

Plato believed the key to understanding nature lay in seeking the reality that underlies what the eye perceives. So he focused on the workings of the underlying details of planetary motion. He argued in *The Republic* that by exploring things in detail one uncovers a much simpler, constant, and more intellectually satisfying truth. Plato sought the unchanging ideal of geometrical truth in the circle. Why a circle? Mathematically, anything that moves in a circle does so without end, and it keeps the same distance from the center. This is exactly what the stars did. What subjects beyond those in the highest of heavens could better be suited to point the way to eternal truth? And if the planets do not seem to execute perfect circular motions, then we must enter the house of geometry to construct combinations of tracks that would cause them to do so.

It is easy to see where Eudoxus, one of Plato's pupils who was challenged by his mentor to work out the geometrical details of planetary orbits, got his idea. The cosmos according to Plato resembled a spinning top made up of eight concentric shells rotating around the same shaft at slightly different speeds. Rather than exhorting a prime mover to spin his top, Plato resorted to allegorical figures like Necessity, the Daughters of Necessity, and the Fates. For example, he tells us that the axle turns on the knees of Necessity and that each shell is a Siren who gives out sounds of a certain pitch to indicate the speed with which she turns. All Siren sounds blend together to create the sublime music of the spheres.

Or the spheres can be offset. Ever the clever geometers, the Greeks also realized that they could resort to this technique as a way of accounting even more precisely for the fact that the planets as seen in the sky do not move uniformly. Alexandrian astronomer Ptolemy (ca. AD 150) supplanted Eudoxus's model with a scheme that called for each planet to travel on an orb (epicycle), the center of which traveled on a larger orb (deferent) around a point slightly off center from the earth. Although this explanation might not seem aesthetically pleasing to us—whoever heard of something going around a mathematical point?—we must remember that the common sense behind Greek models of the universe lay in saving the appearances by using circular motion. Besides, the Greeks had never heard of gravitation. So why should we penalize them for being unaware of our hindsight?

We all share in a deep-seated human urge to replicate the world around us—to model it. Such a desire is represented in the first prehistoric rock art and it continues today in computerized sports, theme parks, and reality shows. (See Chapter 6, which is on clocks—our own way of replicating time.) Before the mechanical clock came the water clock, before that the sundial, the quantified descendant of a simple stick placed straight up in the ground or the shadow cast by a person's arm pointed skyward.

The Greeks were no different. They called their versions of the real world *simulacra,* a word meaning "to simulate": they simulated gods, animals that walked, birds that flew, ships that sailed—even whole battle scenes. Automatons in general were particularly popular throughout the classical world. The Egyptians devised talking statues of the sun god, Ra. In ancient Rome, as a way of arousing public action against the conspirators, Marc Antony was said to have devised a talking wax automaton of Julius Caesar to accompany his funeral oration of the assassinated emperor. As Antony spoke, the statue, operated by a mechanism beneath the bier, rose from a mock coffin and exposed the still bleeding twenty-three knife wounds. Horrified onlookers were said to have been moved to vengeance.[6]

The Tower of the Winds adjacent to the old Athenian agora was a veritable storehouse of cosmological knowledge and information conveyed via *simulacra.* At one time it housed a huge water clock, several sundials, and a weather vane. The Antikythera mechanism, found in a shipwreck off the coast of Asia Minor and dated to 80 BC, was another *simulacrum,* a mechanical version of the movement of the sun and moon among the stars, all played out on a sophisticated system of interlocking gears. These models showing how things move in the universe were created for purely aesthetic purposes, not so much to give precise renditions of nature's wonders but rather to glorify them. Concrete and real though they may be—as real as a planetarium or my cardboard and clay model of the solar system—we should never lose sight of the fact that models are really based on ideas; they all emerge from notions conceived in the mind. Remember this the next time you visit your local science center.

Among the earliest and most imaginative Greek cosmological models was that of Ionian philosopher Anaximander of Miletus (611–547 BC). A mapmaker by trade, he conceived of the earth as a hollow cylinder, one-third as deep as it is wide (which, much to my surprise, is replicated precisely in a standard hockey puck, whose dimensions are 3" x 1"). If you look at the extent of the ancient Greek world and the means of getting to its periphery, a hockey puck model makes sense. The Greek sphere of influence in the fifth century BC included the Aegean, which practically surrounded the roundish Peloponnesian peninsula. The Aegean was surrounded by distant lands, which in turn were encircled by a vast ocean. The Greeks knew of the western bounds of the world beyond Gibraltar and they had some knowledge that a large body of water (the Indian Ocean) lay to the east. Ever devotees of the circle, they had little trouble imagining that a circular ocean dividing earth and sky lay at the disk's rim. We live on one surface of the cylinder, speculated the imaginative Anaximander. Above us lie wheels of fire; but we are protected from them by a sphere of air. The stars are tiny "heating holes" that puncture the air sphere allowing us to see the fire beyond. The sun and moon are larger apertures, the latter a variable slot (think of the adjustable lens of a camera), which explains why the moon waxes and wanes through its phases. Earth remains fixed in the center, equidistant from all points on the sky.

How did the hockey puck universe originate? According to Anaximander, the material basis of all things was *apeiron*—the unbounded, the unrestricted, the unlimited—an indefinable essence capable of taking on all possible forms. The world arrived at its present state when, somehow in the eternal movement that characterized it, it became differentiated—separated out into cold earth and a surrounding hot ring of fire. This indefinable *apeiron* has often been confused with air. The philosopher Anaxagoras, a contemporary of Anaximander, actually argued that air was the substance from which the entire universe was created.

The shift in emphasis toward orbs and away from stacked layers was a Greek phenomenon arising in part from their contemplation of the social universe, which then gave rise to the same sort of intellectualizing about the physical universe. French classicist Jean-Pierre Vernant notes that in stark contrast to its Babylonian predecessors, Greek astronomy had little to do with religion. Anaximander's world view mentions not a single deity. It offers instead a rational process for keeping the earth motionless at the center of things. The levels of mythic space in other stacked universes, each with its own composition, content, and purpose, are gone, replaced with a homogeneous space differentiated by places and the distances between them—geometrical space.

As Vernant notes, you can't take Anaximander's ideas out of historical context and simply regard his as another one of those clever models the geometry-loving Greeks churned out. For if you do you miss the fact that Ionian philosophy was developed during a time of radical economic and social transformation in the Greek city-state. If you examine it carefully, you will note that the language Anaximander uses to articulate his virtual universe really sounds as if it could have come from a modern political science text. He tells us, for example, that the earth stands motionless because it lies at the mathematical center of all things in perfect equilibrium. Nothing holds power over the earth; nothing dominates it. It need not rest on anything.

Many early Greek philosophers referred to the earth as Hestia, the hearth of the universe, once again structuring the natural realm in accordance with the condition of the social world. In the Greek world, nothing dominated the agora, the public power point where each citizen could take his place and speak freely. This idea stood in stark contrast to the hierarchical domination of the king over successive lower classes in the more or less pyramidal social arrangement that preceded the Greek city-state. Likewise, the organization of planets into geometrically measured orbs that go round and round like clockwork was a long way from the stacked universes of the Aztecs and the Navajos—universes worth probing in their own right, but not the one bequeathed to the modern West.

Geometry—from *ge* ("earth") and *metria* ("measure") or *metrein* ("to measure")—is a Greek word. It means "the measurement of the earth" and, as we have seen, the Greeks loved it. But the abstract geometry we learned in high school had a far more practical purpose in ancient Greece. As its root implies, geometry was used to measure land. By the seventh century BC, when they first began to build cities, the Greeks had developed a fascination with the organization of things in space. Surveying and city planning were among the tasks that occupied early Greek geometers. In his play *The Birds,* written around 400 BC, Aristophanes includes a scene that ridicules a city planner named Meton. Intended as comic relief, the scene opens with the surveyor stumbling out onto the stage, his arms loaded down with instruments:

> METON: I come to land-survey this Air of yours. And mete it out by acres.
>
> PISTHETAIROS: Heaven and Earth: Whoever are you?
>
> METON: (*scandalized*) Whoever am I! I'm Meton. Known throughout Hellas and Colonus.
>
> PISTHETAIROS: Aye. And what are *these?*
>
> METON: They're rods for Air-surveying. I'll just explain. The Air's, in out-line, like One Vast extinguisher; so then, observe Applying here my flexible rod, and fixing My compass there—you understand?
>
> PISTHETAIROS: I don't.
>
> METON: With the straight rod I measure out, so that The circle may be squared; and in the centre A market-place; and streets be leading to it Straight to the very centre; just as from A star, though circular, straight rays flash out In all directions.
>
> PISTHETAIROS: Why, the man's a Thales![7]

If this abstract scene is incomprehensible to a contemporary reader, imagine what it must have sounded like to an ancient Greek theatergoer who was not familiar with the latest theories of the universe. Just as pop culture might mock a modern physicist for conjuring up string universes and invisible quarks—"It's all Greek to me!"—so too the average Athenian citizen might have had a hard time appreciating the likes of Anaximander or Meton. But given what we now know about Anaximander's model, the satire in the passage from Aristophanes (especially the opening lines about Air) should make sense. Meton, by the way, actually was better known as an astronomer. He was the first to recognize the Metonic cycle, or the time it takes a full moon to return to the same date in the seasonal year: 19 years or 235 lunar phase cycles, to be exact. Meton's plan, while the object of satire, actually may have been put into practice by a contemporary town planner, Hippodamus of Miletus. He rebuilt his own town on the island of Rhodes after it had been destroyed and he also planned Athens's port city of Piraeus. His reconstruction of Miletus featured a rational geometrical model. After chosing an open, flat space, Hippodamus constructed a right-angle grid of streets around it so that the placement of the buildings in urban space would make for the most efficient political relations.

The battle over the role social mores and customs play in the direction and conduct of science will forever be waged. Every point always has its counterpoint, and although some scholars (like Vernant) claim that it was the new Greek economy and politics that directed them toward geometrical speculation, others read the record differently. Historians Stephen Toulmin and June Goodfield, for example, think the Aristophanes passage from *The Birds* is really intended to poke fun at an isolated, out of touch, minority group consisting of men of leisure, little affected by the workings of the state—men who had little else to do but speculate on nature and the universe.[8] Anaxagoras, they point out, was so unpopular that he was imprisoned for his teachings about nature and later expelled from Athens. Consequently, the fruits of his theoretical labors fell upon the ears of but a few hundred disciples. Still no one on either side of this ideological debate will deny that there is a connection between how people live and what they grasp about the universe around them.

My students have great difficulty appreciating that our Western obsession with geometry—Pythagorean triangles, the golden mean, and all the theorems drummed into our math-resistant adolescent heads—is a direct result of our descent from the Greeks. A concerned student once protested, You mean everyone on earth doesn't believe these things? Connecting astronomy and city planning is a bit like asking one of my young charges to search out theories of cosmology in a textbook on political science or economics; but the connection is important historically, for it is exactly the kind of analogy we need to explore to find out where our contemporary common sense about a universe consisting of things in orbits came from. Aside from reflecting the dual interests of astronomy and city planning, there are some telling clues in the passage from Aristophanes that directly link our contemporary model of the solar system with that of the ancient Greek city. Take the highly favored idea of everything focusing on the center of a circle, like the converging rays of the sun. The urban design of Washington, D.C., is a modern example.

In *The Politics,* Aristotle (fourth century BC) weighs in on how to construct the ideal democratic city, the *polis.* His very first word is one that we associate today with astronomy:

> Revolving round the market-place and the city-centre, people of this class [the shipkeepers, mechanics, and day laborers] generally find it easy to attend the sessions of the popular assembly—unlike the farmers who, scattered through the country-side, neither meet so often nor feel so much the need for society of this sort.[9]

The agora—that dynamic center of motion, the mathematical focal point of the city of Athens (today we might call it "the commons" or "the village green") would become the sun; and each of the various social groups that contribute to the life of the city would, like the planets, occupy their appropriate spheres of action about it.

In the Greek democracy, everything lay in the public domain; all matters were subject to open debate and argumentation in the agora. When it came to decision making on civic matters, the priest-king in the hierarchy of power was replaced by the citizens arranged in perfect equilibrium. Theirs was a balanced system. Running the city came to be more a secular than a religious activity. Such a breakthrough in the design of the ordered universe came about, as it did in all the other models we've looked at, as a result of everyday, real life experience.

So can we really claim that our universe—the one that consists of planets orbiting the sun, the sun orbiting the Milky Way galaxy, and our galaxy orbiting other galaxies—is *the* one true universe? A radical reductionist might answer yes. But in my view it would be far more accurate historically to say that our present orbital concept of the universe descended from a Greek idea based on the organization of people in the city, which in turn gave birth to a similar idea about the arrangement of planets in space. Interesting word, *idea.* It comes from the Greek word *idia,* meaning the look or appearance of something that you can see, as opposed to what it really is. Gus was right: it is Greek!

NOTES

1. Stephen Toulmin and June Goodfield, *The Fabric of the Heavens: The Development of Astronomy and Dynamics* (New York: Harper, 1961), 79.

2. Jean-Pierre Vernant, *Myth and Thought Among the Greeks* (London: Routledge, 1983), 185.

3. Paul Zolbrod, ed. and tr., *Diné Bahané: The Navajo Creation Story* (Albuquerque: University of New Mexico Press, 1984), 41.

4. Derek de Solla Price, *Science Since Babylon* (New Haven: Yale University Press, 1975), 54.

5. Simplicius, in *Aristotelis Quatuor Libros de Caelo Commentaria,* cited in Pierre Duhem, *To Save the Phenomena: An Essay on the Idea of Physical Theory from Plato to Galileo* (Chicago: University Chicago Press, 1969), 5.

6. Price, *Science Since Babylon,* 52.

7. Benjamin Rogers, ed., *Aristophanes in Three Volumes: 3. The Birds* (Cambridge: Harvard University Press, 1924).

8. Stephen Toulmin and June Goodfield, *The Discovery of Time* (Chicago: University of Chicago Press, 1965).

9. Ernest Barker, ed. and tr., *The Politics of Aristotle* (Oxford: Oxford University Press, 1946), 265.

THREE

WHY YOU CAN'T HAVE A BIG BANG WITHOUT DARWIN—AND VICE VERSA

Creation is unbelievable to an atheist. But it is believable to those who believe in God.

—DUANE GISH, DIRECTOR OF THE INSTITUTE OF CREATION RESEARCH[1]

The spectacular perfection of that nest, that tiny tongue, that beak calibrated perfectly to the length of the tubular red flowers from which she sucks nectar and takes away pollen to commit the essential act of copulation for the plant that feeds her—every piece of this thing and all of it, my God. You might be expressing your reverence for the details of a world created in seven days, 4,004 years ago (according to some biblical calculations), by a divine being approximately human in shape. Or you might be revering the details of a world created by a billion years of natural selection acting utterly without fail on every single life-form, one life at a time. For my money the latter is the greatest show on earth, and a church service to end all. I have never understood how anyone could have the slightest trouble blending religious awe with a full comprehension of the workings of life's creation.

—AUTHOR BARBARA KINGSOLVER, "A FIST IN THE EYE OF GOD"[2]

The diversity of life on earth is the outcome of evolution: an unsupervised, impersonal, unpredictable and natural process of temporal descent with genetic modification that is affected by natural selection, chance, historical contingencies and changing environments.

—THE AMERICAN ASSOCIATION OF BIOLOGY TEACHERS (1995)[3]

I find it hard to accept the Big Bang theory. I would like to reject it; it is such a strange conclusion. . . . [I]t cannot really be true.

—COSMOLOGIST ALLAN SANDAGE[4]

Hands down, the most illuminating moment in my five decades in the classroom happened about twenty years ago. I was team-teaching a course titled "Evolution" with a biologist and a social historian. As the resident physical scientist, my task paralleled the one I have set forth in this chapter: to deal with how Darwin's view of life influenced ideas about the inanimate world and vice versa. On this particular day the biologist, a giant of a man with a booming voice, was holding forth in the pit of the lecture hall on the subject of DNA. Backed by abstract-looking blackboard diagrams, the language of his presentation was heavily laced with science jargon: "And then the messenger RNA . . ."; "Along the network of neurons . . ."; "The coding segments of the gene . . ." As the lecture proceeded, the social historian, a passionate man noted for his forceful candor (often shared without solicitation), began to grimace and squirm in his chair.

Continued the biologist: "Once this information is received . . ." and "If the switch is turned on . . .", as the historian's grimace changed to red-faced apoplexy, his squirm to a discomforting writhing. Suddenly he lurched out of his back-row seat next to me and shouted over his pointing finger: "Wait a minute! Wait a minute!" Startled, the bulky biologist dropped his chalk and wheeled his large frame around like a radar dish attempting to hone in on the direction of the disturbance. "You're talking about DNA as if it were a computer!" hollered the historian. After a long pause the life scientist retorted, "Well it is! (another long pause) Isn't it?" Several dozen gaping young minds swiveled back and forth to follow the impassioned dialog. "The words you're using—network, receptor—you're telling your audience that a molecule is a computer!" repeated the historian. Countered the biologist: "Well, if it walks like a duck and quacks like a duck, then it *is* a duck!" The two academes argued tit-for-tat for the remainder of the period and beyond. And I witnessed one of the great unanticipated, unrehearsed—and, I would argue, most important—events in my science classroom experience.

My memorable anecdote cuts to the core of a central question about the way we think science works—or doesn't. Are the models we create approximations to unattainable truths—mere extensions of our intellect—or do they unmask a real hidden truth? Do our explanations reveal the real thing or are they only metaphors—the cobbling together of tangible, familiar, and contemporary shapes, sights, and sounds that offer tantalizing tastes of unreachable underlying essences that constitute the real world? As we have seen, the Greeks had their answers to those questions. Our answers may be different.

Metaphor or reality? The question seems unanswerable. Those who engage it agree that there's a sense of satisfaction when analogies—of waves lapping on a shore for electromagnetic forces, falling bodies for orbiting bodies, or orbiting planets for orbiting electrons—succeed in neatly predicting measurable, certifiable outcomes. Applied to the classroom outburst described above, the question becomes, is DNA really a computer or is the computer just a familiar popular device whose parts and processes so preoccupy us that we can't resist conveniently making use of them to shed light on how life works at a microcosmic level? If you think about it, a piece of flesh and a calculating machine are an unlikely pair of images to meet on the common ground of discovery. But then so were the moon and Newton's apple. Let me provide another example to illustrate my point that a concept from one end of the continuum of scientific perspective can resonate in another.

What an exciting time it was in science at the turn of the twentieth century. In 1896, within two months of the discovery of X rays, came the revelation of radioactivity. French physicist Henri Becquerel wrapped a photographic plate in black paper, placed it on a layer of a ground-up uranium salt, and laid a coin on top of it. When he developed the plate, out came the image of the coin. It seemed an act of magic. The uranium salt had emitted its own radiation spontaneously. Within two years Marie Curie recognized the even more powerful radioactive properties of thorium, polonium, and radium—and later paid the price for her curiosity with her life when she died a result of radiation poisoning. Here were atoms that were not unchangeable and indestructible. They emanated alpha and beta particles and gamma rays as they spontaneously transformed themselves into other atomic species—more magic!

In the immediate aftermath of these events hardly a month passed without some news of the discovery and use of radioactive phenomena. Radioactive dating became the dominant method for pinning down the age of the earth, and radioactivity was recognized as one of the major sources of energy in the earth's interior. Eventually, harnessing the power of the decaying atom would pass well beyond the A-bomb into the realm of medical therapy. Mysterious, potentially dangerous and destructive denizens of nature became saviors of human life. Textbooks of the 1930s and 1940s heralded radioactivity as a major force in the universe. It was everywhere in the public sector, and where it wasn't, it was actively sought out.

Given these developments, I doubt it was mere coincidence that in 1927 in one of the astrophysical journals a Belgian mathematician published an article that would play a direct role in launching one of the great theories of cosmogony (the study of the origin of the universe)—the Big Bang theory. The man was Georges Lemaître, also an abbé (a member of the secular clergy); the title, "The Primeval Atom." That Lemaître had seized the day is evident in the opening sentences of his now historic paper in which he hypothesized that the creation of the entire universe happened when a single, gigantic, radioactive atom spontaneously decayed:

> I was led to formulate this hypothesis, some fifteen years ago, from thermodynamic considerations while trying to interpret the law of degradation of energy in the frame of quantum theory. Since then, the discovery of the universality of radioactivity shown by artificially provoked disintegrations, as well as the establishment of the corpuscular nature of cosmic rays, manifested by the force which the Earth's magnetic field exercises on these rays, made more plausible an hypothesis which assigned a radioactive origin to these rays, as well as to all existing matter.[5]

Lemaître's primeval atom, a concentration of all the visible matter and energy of the universe, filled a very small space, not much bigger than the size of the earth's orbit. It existed only for an instant. Because it was unstable, as soon as it came into being it broke into pieces that were further shattered into protons, electrons, and other particles. The whole of it expanded outward to fill all of space as we see it today. The biggest atomic chunks, like uranium, continue to give us a sampling of the disintegration process.

This exploding chaos of cosmic billiard balls may sound naïve to us, but Lemaître had unwittingly sown the seeds of the Big Bang theory, the idea that we live in an expanding, ever changing universe, with clues to its history buried in its present state—clues that reflect conditions once quite different from what we see around us in the cosmic environment today. An overwhelming majority of cosmologists believe the condition of the universe today is the direct result of events that happened long ago—for we live in an evolutionary universe.

"The times they are a-changin'." "The world is changing." "Change is good." "Change is progress." "Change is for the better." "Let's all work for change." There is scarcely a course in my university's catalog that doesn't purport to deal with change: Child Adolescent Development, Colonialism and Development, Cultural Continuity and Change, Developmental Psychology, Economic Development, International Development, and Moral Development and Education; not to mention a host of courses implying change: Art Since 1950, Early American Literature, Early Twentieth Century Art, and Europe Since 1815.

Evolution is about change. It comes from the Latin *evolutio,* which means "unrolling"—the way you'd read an ancient papyrus scroll. To evolve means to unfold, to reveal parts already present but in a compacted form, out of sight—the way an embryo unfolds to make a person. At least that's how the word was first used. Today, biologists who talk about evolution don't mean to imply that new species can't be created because all components of life were locked in place when life originated. They have altered the meaning of the term. But make no mistake, evolution is a theory about development, about the way things change.

Evolution really began with Aristotle. Like Darwin, he paid quite a bit of attention to living organisms, from swarming bees to hatching chicks. That he saw historical continuity in living forms is directly reflected in what he wrote about life:

> The transition by which Nature passes from lifeless things to animal life is so insensible that one can determine no exact line of demarcation, nor say for certain on which side any intermediate form should lie. As one goes up the ladder, next after lifeless things comes the class of vegetables, and these display quite different amounts of vitality; so that, in short, the whole vegetable kingdom, while lacking "life" if compared with animals, possesses a great deal of "life" by comparison with other bodies. Among plants then (as we said) a continuous scale of forms is observed leading up to the animal kingdom. . . .
>
> Thus Nature passes from lifeless things to animals in an unbroken sequence, through a range of intermediate beings which are alive and yet do not represent true "animals"; and so close together are all these neighbouring groups of creatures that there seems scarcely to be any clear distinction between them.[6]

We can glimpse the seeds of Darwin's ideas in the way Aristotle describes the "ladder of nature." He almost seems to be saying that there is a time sequence to the various forms that make up nature's tree of life. But that interpretation is based on a hindsight reading of phrases like "passes from lifeless things to animals." Aristotle was, in fact, more interested in the variety of living forms and how they functioned than in proposing how species underwent change. The ladder of life he envisaged was static. He believed in an eternity of natural forms created out of necessity: Rain doesn't fall in order to mature our crops. Our crops are nurtured because the rain happens to fall. Likewise, we have sharp teeth so that we can chew meat.

Just as it took over the planets (see Chapter 1), Christianity also would absorb these pagan concepts of nature and convert the universe into a different sort of static aggregate consisting of symbolic works that manifested the glory of God. The plan for all of God's creatures was thought to be fixed and immutable. To observational science went the job of collecting data to demonstrate the perfection of God's creation (natural theology) until Enlightenment philosophers began to question the purposes underlying the quest for such knowledge. Contrast botanist John Ray's views on descent (1691) as a manifestation of the extraordinary creative power of the deity, with Maupertuis's naturalistic point of view, as he discusses reproduction in a work written half a century later in 1745:

> What can we infer from all this? If the number of Creatures be so exceeding [*sic*] great, how great, nay, immense, must needs be the Power and Wisdom of him who form'd them all! . . . As it argues and manifests more Skill by far in an Artificer, to be able to frame both *Clocks* and *Watches,* and *Pumps,* and *Mills,* and *Granadoes,* and *Rockets,* than he could display in making but one of those sorts of engines; so the Almighty discovers more of his Wisdom in forming such a vast multitude of different sorts of Creatures, and all with admirable and irreprovable Art, than if he had created but a few; for this declares the greatness and unbounded capacity of his Understanding.[7]

> Why should not a cohesive force, if it exists in Nature, have a role in the formation of animal bodies? If there are, in each of the seminal seeds, particles predetermined to form the heart, the head, the entrails, the arms and the legs, if these particular particles had a special attraction for those which are to be their immediate neighbors in the animal body, this would lead to the formation of the fetus.[8]

A major issue surrounding evolution has always been whether the changes we see are directed from within by an "arrow of time" that points us in a certain direction (evolution by design) or controlled from without by our responses to stimuli in a changing environment (environmental determinism). A second issue pivots around the individual versus the group. Do changes that affect an individual become part of the evolutionary process or is it only species that evolve? Darwin's innovative point of view was that there are great variations among the offspring of all individuals and that generations of survivors manage to succeed because they possess the advantage of a particular trait or traits over those that do not succeed. Furthermore, the sum of those traits best keyed to survival will define the makeup of future generations of individuals. What makes the game more intriguing is that the conditions that make for survival also change. Viewed from its ends, evolution is a way of eliminating the unfit. It is purposeless, wasteful, death-filled, and goalless; and it is based entirely on trial and error—a pure unadulterated crapshoot. Darwin says no to the sanctity of the individual and no to time's arrow. No wonder he raises our hackles!

Others see a firm hand directing time's arrow. We are already familiar with the idea of intelligent design. We touched on it in Chapter 1 when we discussed the doctrine of astrology, which throughout history has held that our lives are directed from without, that general patterns of behavior or certain tendencies are preprogrammed into your future life. Face it; it's a duck! Intelligent designers avoid the messiness of Darwinism by taking what they see at the cellular level at face value. As Michael Behe, one of intelligent design's foremost proponents, characterizes it:

> In the past 50 years modern science has shown that the cell, the very foundation of life, is run by machines made of molecules. There are little molecular trucks in the cell to ferry supplies, little outboard motors to push a cell through liquid. . . . [T]he entire cell can be viewed as a factory with an elaborate network of interlocking assembly lines, each of which is composed of a set of large protein machines.[9]

Machine is not metaphor. This term is applied literally. And machines as sophisticated as this are not created accidentally.

Unlike the determinists, the designers see evolution as a creative process, a way of creating the fit by gradually building up a pool of survivors, the way many of Darwin's contemporaries saw it. If evolution by design is true, then all species, like the porridge that gets eaten in the story of Goldilocks and the three bears, are constructed to permit them to function just right in the environment. The intelligent design argument takes the position that there is both pattern and purpose to life. The laws of nature, rather than acting blindly and without cause, really function to change species in such a way that they can survive. Intelligent designers seem as concerned with the search for meaning as do theologians contemplating the mystery of the Star of Bethlehem. This extraordinary polarization of views is bound to elicit great passion, as indicated by my opening epigraphs.

In Darwin's day there was a way to take the sting out of the ruthlessness of evolutionary selection: the theory of progress. Even if it did involve random processes, if evolution could be taken to imply an increase in complexity characterized by the specialization of functions and the integration of these functions into higher levels of organization, then maybe evolution was endowed with a built-in arrow of time—an arrow that pointed uphill. After all, this was the way contemporaries of Darwin characterized the progressive development of all societies—from small, simple, and low tech to the large, modern, high tech, and industrial. Darwin's idea was strongly influenced by the prevailing nineteenth-century viewpoint that capitalism was the most successful strategy for a progressive society: you define the market as a struggle for individual profit and, in the principle of survival, the edge goes to the most successfully competitive—and this leads to progress. Division of labor and the acquisition of more specialized skills were among those characteristics believed to chart the directional development of all human cultures, from savagery to barbarism to high civilization. Today anthropologists realize that such evolutionary schemes are highly ethnocentric, for they have a tendency to place contemporary Western political and religious systems highest on the slope of human development.

On the scientific side, it took nearly a century after the publication of *The Origin of Species* (1859) for Darwin's theory of natural selection to incorporate new studies on mutations, population genetics, and the paleontological record and to reemerge as we largely recognize it today; that is, as a theory that relegates practically all change in species and subspecies to adaptation that happens by chance largely at the genetic level. If anything, the purpose of life is to acquire a partner in sex so that you can propagate your genes.

Ultra-Darwinism is an extreme reductionist school of thought that virtually eliminates any vestige of creative design elements in life by taking the organism out of the picture entirely, advocating instead that the "selfish gene" (the term "selfish" in this sense being used metaphorically) does all the selecting irrespective of the individual who carries it. An arch proponent of this view, biologist Richard Dawkins, explains why a bird puts itself at risk by uttering a call of alarm to warn its flock of the impending attack of a hawk: hearing the sound, the hawk attacks the bird, but the flock is spared. Sounds like altruistic behavior, right? But, Dawkins explains, the gene that carries the bird alarm is preserved, not by the bird that gets attacked (obviously), but instead by a multitude of surviving relatives in the flock who all carry similar genes, including that of the alarm. At a more personal level, suppose my younger brother, who is actively engaged in raising a family, is dying of kidney disease and I (say, unmarried and highly moral) offer to donate one of my kidneys to him. You might say my action is altruistic. Suppose further that such a sacrifice were made at the risk of making me impotent so that my brother might continue the ability to procreate; isn't that altruism in the highest sense? As far as the genes are concerned, my act, by prolonging my brother's ability to procreate, passes on the same genes. And so evolution by selection proceeds, but it is the genes and not the organism that engage in life's great struggle. I am no more than a machine that replicates DNA.

The social, political, and moral ramifications of natural selection are with us now as they were over a century ago. If my abilities are prescribed by my genes, perhaps so are those of my race or ethnic group. And if any defect in my genetic strain can be isolated, maybe it also can be eliminated to enhance the survivability of my future genetic pool. Be selfish and survive! Few will deny that, in the 150 years that the ever-changing concept of evolution has survived, it has had a tendency to move us closer to a purposeless, trial-and-error existence—a world view in which everything from the structure of the human brain to the cosmos just happens.

My dentist tells the following story. A woman brings her teenage son in to treat an emerging wisdom tooth that has become inflamed. Engaging in the usual, casual, one-way conversation with his open-mouthed patients, the doctor drops the remark to his young victim that these particular teeth of his are largely useless relics of a bygone age when our ancestors required more grinding power. Eons from now, he continues, wisdom teeth will probably diminish, the way the tail has shrunken to become the appendix, until they vanish altogether from the human organism. "You mean to tell me you believe all that evolution stuff?" replies the mother in shock. "How can you swallow such horrible rubbish?" Next day she abruptly cancels all future appointments and asks to have her son's file sent to another office.

When an acquired truth bonds a community, there's a reluctance to relinquish its patterns of thought. A Gallup poll taken at the end of 2004 revealed that 55 percent of all Americans believed that less than 10,000 years ago God created the universe just as we find it today. Another 27 percent believed that we evolved from less advanced forms over millions of years, but that God directed the process; only 13 percent—just a little over one in ten—bought Darwin's argument.[10] Even if they can allow for an interpretation of the Genesis seven-day time scale of creation to include geological eons, the devoutly religious cannot imagine such a wasteful deity who would eliminate all those who offer no benefit to the species. Science may have progressed, but in the public domain evolution by selection, at least in the eye of the American public, has not persuaded many to abandon the traditional creationist beliefs.

Does this mean science and religion conflict? Or do they only offer us different kinds of truth? Consider two hypothetical job descriptions:

(1) WANTED: Full-time worker to explain natural phenomena in terms of laws inherent in nature and discoverable through experiment. Arguments must be rigorously and precisely formulated and proceed by logic. Special qualifications: total faith in the existence and discoverability of such laws; complete objectivity; ability to function dispassionately; willingness to run risks, which include deliberately contriving hypotheses that can lead to predictions that may either falsify or advance their hypothesis; quantitative skills highly desirable.

(2) WANTED: Open-minded, expansive person willing to work to seek human understanding and meaning in the perceived state of the world, both as it applies to human culture and to nature; required to fashion a set of moral and ethical principles consistent with the pursuit of a felicitous existence in which all fellow humans have a participatory role. Special qualifications: tolerance; compassion; openness to alternative points of view based on moral principles; faith in the existence of a purposive universe acquired from revealed knowledge.

Do you see conflict in the day-to-day jobs of scientific versus religiously minded individuals? Science may offer us its truth about how nature operates. It is dynamic, rule bound, testable, quantifiable, and precise. Religion is about truth that deals with the search for meaning. What it offers us has to do with building community, belonging, and being content.

But some say there is infringement. They argue that since the Enlightenment, science has encroached on questions traditionally the purview of religion and that it has proven the religious view of the creation of the world to be just plain wrong. Early pro-Darwinists—an especially virulent group was operating in Scotland in the 1880s—were quite open about their ends: to secularize all of society using evolution as their principal means. They waged an all-out war against opponents who would allow supernatural occurrences to be a part of scientific explanation. They charged themselves with taking cosmology entirely out of the realm of theology: "All schemes and systems which thus infringe upon the domain of science must, in so far as they do this, submit to its control and relinquish all thought of controlling it."[11] Today science still secularizes and, even if it doesn't attack religion directly, a cold war exists between the two. Although some would say that no scientific discovery can directly imply ethical conclusions, the 1995 statement of the American National Association of Biology Teachers on evolution (the third epigraph in this chapter) leaves little wiggle room for a supernatural creator.

So do the twain meet? Maybe unmasking truths about human origins ought to be left to the hard evidence of the sciences and queries into life's meaning and moral conduct should reside in the realm of religion. My own view (as mentioned in Chapter 1) is that it is difficult to deny that religion and science—two radically different ways of knowing—abrade one another, especially when one extreme branch of materialistic scientists attempts to reduce even our morals to the genetic level, while theories emanating from another equally extreme faction on the side of intelligent design (the scientific creationists) masquerade religion in scientific method.

Having explored the debate over how and why we believe living things change, let's turn back to the inanimate universe. Like flesh and the calculating machine, people and things also come together on the common ground of discovery. If people change then why not things? Why not the entire universe? A universe that changes is common sense today, but the idea that matter has a history of its own didn't develop until the late eighteenth century. That's when naturalists first challenged the static view of the world based on biblical literature, which posited that all things reflect an unchanging grand design laid down by a divine creator. That version of the universe began in an instant and it will end in an instant with the destruction of the world on judgment day. For the literal-minded, the material world was useless except in relation to animate intellectual beings. It was all created solely for the purpose of contemplation—to frame one's moral character via symbols of the perfection of a great and wise creator. But careful observation of blemishes on the face of the earth revealed that nature had already begun to unravel a timeline—rivers eroded the landscape, wind wore down mountains, fossils of exotic plants and animals lay buried below the detritus. As more and more biological species were discovered, these varied species began to look more and more like ancestors and descendants than unrelated creatures created individually out of nothing.

Open any astronomy text and you will find a treasure trove of evolutionary concepts and ideas. Invariably you'll discover a chapter—usually toward the end and usually following chapters on stellar evolution and the evolution of galaxies—titled something along the lines of "Evolution of the Universe." Today all cosmologists agree that the universe has a history. And if the textbook is a good one, its author will at least briefly trace the history of the history of the universe; that is, the changing course of scientific thought about the nature and structure of the universe. All such histories locate the wellspring of a great surge of interest in thoughts about an evolutionary universe in the mid-1930s, which is just about when the new Darwinism, strongly centered around the idea of random adaptation, had begun to coalesce. I did a small survey of astronomy texts going all the way back into the nineteenth century and I can verify that the post–World War II era was the time when the word *evolution* first began to creep into standard literature; for it was also the time when a new theory about the origin of the universe began to gain the upper hand. It was a theory about the ultimate unscrolling—the evolutionary, or Big Bang, theory of the universe.

Stephen Jay Gould has pointed out that biological evolution is variational. Species change because of the differential survival of offspring who adapt best to their particular changing environment. On the other hand cosmic evolution is transformational. Stars evolve according to predictable changes that result from physical laws. Ironically, the cosmological sense of the word *evolution*—an unscrolling or unfolding in a predetermined directed manner—lies closer to the original meaning, even if most of us associate the evolution with Darwin.[12]

Threads of an idea about a universe that grows or develops can actually be traced back to the German philosopher Immanuel Kant fully a century before Darwin. In 1755 Kant wrote: "Matter is bound to certain laws and when it is freely abandoned to these laws it must necessarily bring forth beautiful combinations. It has no freedom to deviate from this perfect plan."[13] Kant was attempting to account for the creation of stars out of the growth of condensations of matter in nebulae (dubbed "shining fluids" by the German-turned-British astronomer Sir William Herschel, who first observed them through a telescope) that appear to surround many stars. Obviously Kant's theory of evolution has more than a hint of God in it, for, as he said, "it must have necessarily been put into such harmonious relationships by a First Cause" ruling over it; and there is a God, because "nature even in chaos cannot proceed otherwise than regularly and according to order."[14] Applied to the solar system, Kant's nebular hypothesis advocated that the sun and its retinue of planets were created quite naturally out of a nebula, a cloud of gas, which, left to itself and the laws of nature—most especially gravity—coalesced to form our solar system.

Few astronomers had read or cared about the words of the philosopher, a sign that natural history was already beginning to break up into the inflexible islands we now call disciplines. (That Kant's publisher went bankrupt didn't help matters.) But when the French scientist Pierre-Simon Marquis de Laplace took over the nebular hypothesis in 1796 by applying rigorous mathematics to the notion that all planets of our solar system grew out of the atmosphere surrounding a contracting, rapidly spinning proto-sun, attention began to be paid. The marquis had the double advantage of the prestige that came from being a mathematician, coupled with good timing. His *Exposition du Systeme du Monde* (1796) appeared at the close of the French Revolution, when his country was wide open to new ideas. Moreover, Laplace was a forceful character. He was fond of making pronouncements like declaring that simply collecting scientific facts is a sterile undertaking. What we must do, he maintained, is theorize about the laws of nature.

Laplace was durable, too. He had a knack for changing ideologies in the face of shifting political winds. He survived the Reign of Terror, Napoleon's empire, and the restoration of the Bourbons (who made him a marquis[15] after Napoleon had titled him Count of the Empire). He was invited by Napoleon to explain his theory, and when the emperor queried, "You have written this huge book on the system of the world without once mentioning the author of the universe," Laplace is said to have responded with the oft quoted statement: "Sire, I had no need of that hypothesis."[16]

In the hands of Laplace, Kant's nebular hypothesis would have a profound effect on scientific, philosophical, and religious thought in the century that followed him. It gave geology a background for the pregeologic history of the earth and it offered the new science a template to account for changes in the earth by gradually accumulating growth processes, an account that directly opposed biblical cataclysm. It also gave biologists a way to view different forms of life as a continuous chain of growth from simple plants and animals to the highest form, man, as expressed in Darwin's *The Origin of Species.*

Not to be lost are the implications of the process of growth and evolution envisioned by Kant, and later Darwin, upon society in general. If nebulae form stars and planets, then we might expect to find myriad inhabited worlds throughout the universe. And if each planet evolves through similar stages of one-dimensional cultural evolution and progressive scientific advancement to the highest form, it makes sense that the universe would be teeming with folks "just like us"—white, male-dominated, European-like societies. I find it interesting that this "they-are-like-us" perspective—although somewhat altered to fit current ideas—still dominates much of the public perception concerning the search for extraterrestrial intelligence.

Listen to ex-SETI (Search for Extraterrestrial Intelligence) director Frank Drake's anticipation regarding what might ensue from an alien encounter:

> There is probably no quicker route to wisdom than to be the student of more-advanced civilizations. . . . I fully expect an alien civilization to bequeath us vast libraries of useful information, to do with as we wish. This "Encyclopedia Galactica" will create the potential for improvements in our lives that we

> cannot predict. . . . I suspect that immortality may be quite common among extraterrestrials. By immortality I mean the indefinite preservation, in a living being, of a growing and continuous set of memories of individual experience. I think this might come about through the development of methods to eliminate the aging process, or to repair indefinitely the damage caused by aging. . . . Some aliens may already know how to transfer immortality from single cells to entire organisms. Or they may be able to transfer the inventory of memories of an old brain into a young brain—perhaps even the brain of a clone, or an exact copy of a being whose individuality is to be preserved. . . . Wars probably don't exist among immortals, who would not take the risk of fighting. I think a civilization of immortals would be extremely active in detecting and communicating with other intelligent civilizations. Not only would they want to extend their resources for amusement beyond their own planetary system, but also communication would grant them the ultimate protection from harm.[17]

Laplace's theory of planetary growth operates on the same set of principles as the scenario proposed by George Gamow and like-minded post–World War II physical evolutionists who put a dramatic spin on Lemaître's long-forgotten idea of a highly compact universe created in an instant eons ago. The key piece of evidence that pointed to an evolving universe was that now there was growing proof that it seems to be expanding. In 1927 (by coincidence, the same year Lemaître proposed his theory of the radioactively exploding primeval atom), Mount Wilson astronomer Edwin Hubble detected redshifts in the spectra of distant galaxies with the then largest 100-inch telescope.[18] Just as the pitch of a sound signal emitted by a receding source, say a blaring car horn, is decreased—which means the wavelength is increased—so is a shift of an electromagnetic signal toward longer wavelengths associated with recession. This huge shift of spectral lines in distant galaxies toward longer wavelengths, when compared with spectra acquired in a fixed earth-based lab, suggested that the galaxies must be moving away from us at colossal speeds.

Furthermore, there was a correlation between the distances of galaxies and their redshifts: the farther away they are from us, the faster they seem to be receding. Evolutionary cosmologists accounted for these observations by running the tape backward, which led to the conclusion that the entire universe had exploded like a bomb, passing from a highly compact existence to its contemporary tenuous state over the several billion intervening years. Curiously, the term "Big Bang" was coined not by Gamow but instead by one of his mocking critics, and it stuck! In strange contrast with Lemaître's theory, which argues that the course of development is a breakdown of a single, complex primordial atom into simpler, less complex constituents, the Big Bang, which assigns high temperatures to the early phases of the universe, argues instead that the expansion and consequent lowering of the temperature was accompanied in the very early stages (in the first half-hour of the expansion) by a buildup of basic atomic structures; for example, neutrons, hydrogen and its isotopes, and some helium were created at this time out of less complex elementary particles.

Like selfish genes, every microstructure that makes up the universe takes on its characteristics in response to the way the environment changes. Neutrons can form only when the ambient temperature and pressure are low enough for electrons and protons to fuse; and they stop forming when the temperature is too low to power their constituents with enough energy to stick together—a matter of minutes in the early phases of expansion. Cool the universe faster or slower and you change the window for neutron formation. The luckiest neutrons—or the "selfish" ones, to borrow Dawkins's metaphor—that bonded with protons to form stable atomic nuclei survived to become the building blocks of the universe we know today. The unlucky ones decayed—in a matter of minutes—into the protons and electrons from whence they came.

Just as dynamic biological evolution had its static counterpoint, the explosive theory of the Big Bang found an opponent in the contemporary Steady State theory. Led by Britain's Astronomer Royal, Sir Fred Hoyle, Steady Staters proffered an extremely rarefied infinite universe uniformly filled for all eternity with galaxies. Yes, the galaxies recede, for one cannot deny the facts of observation; however, for Steady Staters this fact pointed to a different truth: primordial hydrogen constantly comes into existence throughout the universe; it gradually forms into stars, galaxies, and clusters of galaxies in a way that's just right to keep the universe expanding at the observed rate. You might even say that the continuous creation of matter causes the universe to expand.

Cosmologists use a raisin bread analogy to contrast the rival theories of cosmology. In the Big Bang, as the bread that represents the universe bakes (changes), it expands and the space between the raisins (galaxies) increases. The number of raisins per cubic inch of bread diminishes with time as the bread rises in the oven. A same-size slice of bread later in the baking process will contain fewer raisins. On the contrary, according to the Steady State theory new raisins are created as the bread rises, thus keeping the density of raisins the same through both time and space. All slices of the Steady State bread are identical regardless of when you cut it. This condition, expressed in the form of the "perfect cosmological principle," posits that the universe always looks the same everywhere. There is no gross change, no vestige of a beginning and no prospect of an end, as geologists would have said a century earlier concerning the age of the Earth (see Chapter 4 for that story). The Steady State theory was the ultimate uniformitarian challenge to the Big Bang catastrophe. And at mid-twentieth century it polled as many supporters as its evolutionary opponent.

This battle of cosmic ideologies continued until the mid-1960s, when dramatic new evidence swung the tide in favor of the evolutionary point of view. The discovery of quasars, extraordinarily luminous, peculiar galaxies characterized by extremely large redshifts and very active cores, posed a direct challenge to the Steady Staters. If the extreme redshifts of quasars implied extreme distances, that means they must have existed in the distant past (it takes that long for their light to reach us). And that in turn tells us there was at least one characteristic about the universe at large in the distant past (the far away part of the universe) that differs from the universe today (the nearby, local portion). The universe had quasars then and it has none now and that violates the perfect cosmological principle. Bottom line: the universe must have changed.

The 1960s also marked the discovery of the cosmic microwave background radiation. If there had been a catastrophic explosion in the beginning, we would expect the radiation emitted by the Big Bang to thin out and consequently the overall temperature of the universe to drop as the expansion proceeded. If we know the expansion rate of the universe (which we can calculate from data on the redshifts of distant galaxies), then it should be possible to calculate the temperature of the universe today. To give an analogy, suppose I've been preparing dinner in a 400°F oven, which I turn off after dinnertime. Knowing the thermal properties of the oven's insulating material, I ought to be able to predict the temperature inside the oven hours after I shut if off. Big Bang calculations pointed to a contemporary temperature of three degrees above absolute zero. This was confirmed in 1965, when Arno Penzias and Robert W. Wilson, a pair of Bell Telephone Lab engineers, detected a weak signal coming from all directions of space in the microwave part of the spectrum, which is exactly the predicted peak wavelength of such radiation. (They won a Nobel Prize for their efforts.) In the half century since, no one has been able to come up with a viable alternative source for the background radiation, which is why a vast majority of cosmologists remain convinced that the universe has cooled from an initial hot, dense state to its present low temperature, low density condition. In other words, change happened.

Many cosmologists never liked the Steady State idea of continuously creating something out of nothing, but the discovery of quasars and the cosmic microwave background radiation provided the observational evidence for the catastrophists to deal a lethal one-two punch to the cosmic uniformitarians. As is often the case in sports and war, the victors were not kind to the vanquished. At a meeting of cosmologists in the late 1960s, Big Bangers unleashed a barrage of sarcastic poetry on the bowed heads of their fallen foe. Here's my favorite:

THE GHOST OF NEWTON REPLIES

The reason, sir, is simply this;
Boyle and Hoyle we all dismiss:
One's dead,
One's Fred.
All is change, and that's in flux.
Nothing's sure but Lady Luck's
Encrypted program; and who'll compute her
Pressure, volume, vector, mass? What tutor
Taught her that lithe arbitrament
That rules our spacious firmament?[19]

As so often happens in scientific theorizing, out of the ashes of a deflated idea arises a new point of view built along the same lines, but with the capacity to swallow the very notion that struck it down in the first place. The irony here is that the victorious Big Bang may have sown the seeds of a new generation of detractors. Responding to questions regarding the origin of the Big Bang itself, George Gamow once wrote:

> [W]e conclude that our Universe had existed for an eternity of time, that until about five billion years ago it was collapsing uniformly from a state of infinite rarefaction; that five billion years ago it arrived at a state of maximum

compression in which the density of all its matter may have been as great as that of the particles packed in the nucleus of an atom . . . and that the Universe is now on the rebound, dispersing irreversibly toward a state of infinite rarefaction."[20]

Of course, added Gamow, we can never prove it, because we haven't access to any facts before the great compression, what with the entire historical record locked inside an exotic plasma smaller than the head of a pin. At this writing Big Bang theory points to a simultaneous springing into existence at more than one point, like so many black holes operating in reverse. The arrow of time points in the direction of infinite rarefaction. Recent evidence suggests there may be a cosmic force of repulsion, the opposite of gravitation, that actually speeds up the rate of the expansion. This scenario offers us a universe destined to become an infinite dark place more and more tenuously populated by burned out stars—a prognosis as dismal and dispiriting as Darwin's legacy. One astronomy text posits that all of human history is but "a brief and discreditable episode in the life of one of the meaner planets."[21] But I have been around long enough to witness the pendulum swing on this issue a number of times, so I am hedging my bets.

Safe from ever being able to acquire any evidence on the issue, there is nothing lost by supposing, pro or con (as Gamow did) that a big crunch into a black hole had precipitated the Big Bang. Perhaps the history of the universe is really a series of such events: bang, crunch, bang, crunch, ad infinitum. Although not popular at the moment, this theory of the oscillating universe predicts an infinite series of evolutions. Think of the possibilities: because the physical, chemical, geological, and biological processes of each universe are carefully tuned to the environment, any subtle differences in the successive universes created between bang-crunch intervals can produce different percentages and species of atomic and subatomic structures, chemical recombinations, and even living forms. Given enough oscillations, one can imagine some past or future universe morphing a reader just like you browsing a book just like mine, except that you're wearing a different color shirt. (Remember the old monkey-typewriter analogy?) Now isn't there something eerily Steady State–sounding about all of this?

Change encompassed by stasis? What if our inflationary universe is part of a manifold space populated by other inflationary (or deflationary) universes out of contact with one another—separate universes, each with its own history, unaffected and unobservable by us, an infinity of universes forming and dissolving like so many bubbles in a pot of boiling liquid? Maybe rival theories of cosmology are rivals only because of a difference of perspective on the part of the viewer. For example, I recently applied a coat of paint to one of the walls of the room in which I'm now sitting. No matter what portion of that wall I look at, from my vantage point I see pretty much the same uniform, dark yellow monotone. But if I get up and move closer to it and if I take a magnifying glass with me, I begin to discover that my wall is really populated with tiny peaks and valleys—places where the paint is a bit thicker or thinner, where some of the rough underlying structure of plaster beneath becomes visible. The magnifying glass becomes a looking glass, as I discover an entire wall consisting of tiny bumps that span a range of sizes. Backing up again, I witness the close up perspective of chaos and randomness resolved back to a constant monotone.

What if ours is but one of a multitude of universes—a multiverse; and what if only certain of these universes (including ours) possess laws and physical constants tuned to permit life, while other universes are created stillborn by virtue of containing insufficient gravity or the wrong mix of subatomic forces, or perhaps a tendency to collapse a short time after coming into existence? Could a principle of natural selection operate in the multiverse, one that favors the creation of a daughter universe out of surviving parent universes? For example, cosmologist Lee Smolin has speculated that every time a black hole collapses, such a daughter universe (in a different space-time than our own) comes into existence.[22] And what if each daughter universe inherits (perhaps with slight variations) the laws and parameters that governed the parent universe, thus allowing it to propagate other generations of universes in its likeness?

Too many what ifs? Many cosmologists dismiss this biological sort of reasoning as a course embarked upon out of frustration over failed attempts to find a final theory that unites all physical laws. But others have actively pursued the hypothetical engagement with this archipelago of universes, all the way to the possibility of "cosmic genetic tampering"—manufacturing a universe by altering its chemistry and physics.[23]

Intelligent design also finds its way into the physical universe. The Anthropic Cosmological Principle takes two forms.[24] The weak form posits that ours is but one of manifold universes, each with its own set of parameters. No problem. But it departs from natural selection by further proposing that living, intelligent beings could find themselves only in an environment hospitable to their existence. Now, we know that at least one such universe exists. And it is here (get ready for a twist of logic) *because we are in it.* The strong form of the Anthropic Cosmological Principle, analogous to the theological doctrine of biological creationism, alleges straight up that the world was created by a god with exactly the correct physical laws and parameters so that intelligent life could exist.

The words in my fourth epigraph were spoken by an astronomer who, despite his skepticism about the whole idea, made great contributions to the advancement of Big Bang cosmology. A violent beginning and an uncertain end: tough news to digest. This is how the universe evolves. The news is no less deterministic on the biological front. We may alter a few genes and do our best to work for a better environment, but we will never knock evolution off the pathway of adaptive change by chance occurrence. Evolution stirs up our ideological juices precisely because almost any dialog about it is positioned squarely at the frontier between science and human values. What does it have to do with our morals or our ethics; whatever happened to altruistic behavior (remember the story of the bird alarm)? What sort of god would put us in a world like this? Is the future determined by the roll of the dice—by how fast the great expansion took place and by those among us who have the right genetic stuff for survival that we can pass on to the gene pool? Can evolutionary biology really explain our religious behavior? Have we no control over ourselves?

Evolutionary cosmologists don't help their cause by admitting, as they have been forced to do in the past decade, that they have a grip on but 5 percent of the stuff that makes up the universe. Even they admit that the reality they offer us—a picture of galaxies and stars evolving on a complicated mathematical stage of space-time—is likely far from the whole truth.

As a teacher of science I can verify from personal experience that we live in a time of popular revolt against science. In greater numbers than ever students have confided in me their suspicions about science. They don't connect with its rationalist metaphors for the truth. Maybe that's why in that great moment of disagreement that happened in my classroom twenty years ago, the students seemed to be rooting for the social historian. Like the public at large, they fear the quantitative language of science; they find its experimental side tedious and jargon-laden, its discourse too difficult to follow. With everyone looking for a bottom line writ in large type, the public's ignorance about scientific process leads to the unfortunate perception that scientific experts, the high priests of our culture, whimsically shift their positions, especially when it comes to non-esoteric matters: good drugs suddenly turn evil and good cholesterol becomes bad cholesterol, while protein and carbs twice switch poles on the good-evil food axis. Clearly such capricious priests are not to be trusted.

So at the start of a new century we stand divided as we struggle to cope with the idea of change that science has bequeathed us. The community of hard workers inside the fortress of science are having a tough go of it. Imitating the red and blue divide in contemporary American politics, conflicting views polarize us. At stake is not only the public perception of science, but also the more important possibility of the closure of human discourse that comes with making up your mind once and for all about what is the absolute truth.

NOTES

1. Duane Gish, director of the Institute of Creation Research, at a 1975 debate on creationism at the University of Tennessee (© 2004 Institute of Creation Research), www.icr.org.

2. Barbara Kingsolver, "A Fist in the Eye of God," in *Small Wonder: Essays by Barbara Kingsolver* (New York: Harper, 2002), 95.

3. Quoted in Schenectady, New York, *Gazette* (August 15, 2004), F2.

4. Quote attributed to cosmologist Allan Sandage in Gerald Jonas, "The Big Bang and Other Starts," *New York Times Book Review* (November 26, 1978), 216.

5. Georges LeMaître, "The Primeval Atom," in *Theories of the Universe,* ed. Milton Munitz (New York: Free Press, 1957), 339.

6. Aristotle, quoted in Stephen Toulmin and June Goodfield, *The Discovery of Time* (Chicago: University of Chicago Press, 1965), 51.

7. Ibid., 100.

8. T. Broman, "Matter, Force, and the Christian View on the Enlightenment," in *When Science and Christianity Meet,* ed. David Lindberg and Ronald Numbers (Chicago: University of Chicago Press, 2003), 96.

9. Michael Behe, "Design for Living," *New York Times,* Op-Ed (February 7, 2005), A21.

10. CBS News/*New York Times* poll, *New York Times Magazine* (February 20, 2005), 15.

11. Broman, "Matter, Force, and the Christian View on the Enlightenment," 283.

12. Stephen Jay Gould, *I Have Landed: The End of a Beginning of Natural History* (New York: Harmony, 2002), 248–249.

13. Immanuel Kant, "Universal Natural History and the Theory of the Heavens," in *Theories of the Universe,* ed. Milton Munitz, 247n5.

14. Ibid.

15. An honorary title, below Duke, that originally was an appointment to guard the borders (marches or markers) of a kingdom.

16. Quoted in Stephen Toulmin and June Goodfield, *The Fabric of the Heavens: The Development of Astronomy and Dynamics* (New York: Harper, 1961), 254.

17. Frank Drake and Dava Sobel, *Is Anyone Out There? The Scientific Search for Extraterrestrial Intelligence* (New York: Delacorte, 1992).

18. Hubble's discovery actually was preceded by Slipher's little-known discovery made in 1905 at Lowell Observatory in Flagstaff.

19. H. McCord, "Quasi-Stellar Radio Sources," *New York Times* (March 1, 1964).

20. George Gamow, in *Theories of the Universe,* ed. Milton Munitz, 403–404n5.

21. John Charles Duncan, *Astronomy,* 5th ed. (New York: Harper, 1955), 472.

22. Lee Smolin, *The Life of the Cosmos* (New York: Oxford, 1997).

23. These imaginative ideas are reviewed in Martin Rees's *Before the Beginning: Our Universe and Others* (Reading, MA: Helix/Perseus, 1997).

24. John Barrow and Frank Tipler, *The Anthropic Cosmological Principle* (New York: Oxford, 1986).

FOUR

WHEN DISCIPLINES COLLIDE: FROM DINOSAURS TO THE DOGON

Astronomers should leave to astrologers the task of seeking the causes of earthly events in the stars.

—*NEW YORK TIMES* EDITORIAL (1985)[1]

Mathematics may be compared to a mill of exquisite workmanship, which grinds you stuff of any degree of fineness; but nevertheless what you get out depends upon what you put in . . . , so pages of formulae will not get a definite result out of loose data.

—T. H. HUXLEY (1869)[2]

The problem for us therefore is how the Dogon could have known a host of astronomical facts, all of which are invisible to the unaided eye. . . . They have no business knowing any of this.

—ASTRONOMER KEN BRECHER (1979)[3]

Remember all those "ology" courses you were forced to enroll in at college—geology, biology, psychology, anthropology. Ever wonder where these disciplines came from? Who divided up the turf of human inquiry this way and why? Originally codified by fifth- to seventh-century encyclopedists, the liberal arts once numbered seven. The seven pillars of knowledge were divided into two parts: the verbal arts (the *trivium*)—grammar, rhetoric, and logic or dialectic—and the mathematical arts (the *quadrivium*)—arithmetic, music, geometry, and astronomy. This classical canon was transmitted to the middle ages, when the curriculum was redirected from acquiring and transmitting knowledge for its own sake toward the highest of all possible goals—interpreting holy writ.

The scholastic movement sought to use the disciplines to solve problems, such as proving the existence of God and the associated conflicts between faith and reason, will and intellect. The medieval period, however, was also responsible for the structure of the university as we know it today. We hear that word—along with *faculty, master,* and *student*—for the first time about AD 1200. Young men entered an institution that directed them along different pedagogical paths, from law and medicine to arts and clerical. The tremendous outpouring of new ideas and new information in the fifteenth- and sixteenth-century Renaissance demanded an institutional structure for the transmission of new knowledge.

It was only during the great scientific revolution beginning in the early eighteenth century that the study of the natural world began to enter the university as a series of organized disciplines, each with its own distinct subject matter and methodology. Out of medicine grew specialized interests in plants, animals, and minerals; and so the seeds of the disciplines of botany, zoology, and geology were sown. Alchemy was never endorsed in the university, but chemistry emerged, largely as a result of attempts to organize and synthesize salts once it was realized that they could be broken down into more basic chemical elements (behold the power of reductionism!). The oldest of the sciences, astronomy, which centered largely around dynamics (the study of the motion of celestial bodies), took its place alongside physics after experiments in terrestrial laboratories and the invention of the theory of gravitation made it evident that the two shared a theoretical as well as mathematical approach to the material world. Two other "ologies," psychology and archaeology, were twentieth-century latecomers that still reside on the periphery of science. Since then, specialization has further subdivided the scientific disciplines: immunology and epidemiology; behavioral psychology and neuropsychology; organic and inorganic chemistry; sedimentary geology and petrology; planetary and stellar astronomy; nuclear and solid state physics; and so on. Still more recently, disciplinary tools and methods that are shared by more than one discipline have created new *inter*disciplines—geophysics, biophysics, planetary geology, and archaeoastronomy.

As these disciplines have become deeper, more focused, more jargon laden, and more exacting, this increasing specialization has led to enormous progress in many areas, such as understanding how the brain works, what makes atoms tick, how the earth was formed. Yet these advances on the frontier of knowledge have come at the expense of the broad-based knowledge that scientists once had, leaving scientists with little or no understanding of—and a similarly paltry appreciation for—the other disciplines.

As one astute historian observed, "Outsiders tend to see uniformity in other groups and fine distinctions within their own."[4] Consequently, specialization often leads to the point of view that my knowledge trumps yours. A cultural anthropologist has no more comprehension of what evidence derived from the application of universal laws means to a physicist than does a physicist of what evidence derived from the structuralist analysis of myth means to a cultural anthropologist. Let me take you through three case studies in the history of science and show you just what happens when disciplines collide. The examples I've chosen revolve around three basic and quite diverse questions: What killed the dinosaurs? How old is the earth? Could a remote African tribe have acquired extensive knowledge about the stars known to be available only through modern telescopes? These questions span two centuries and they pit the disciplines of physics, astronomy, geology, paleontology, and anthropology against one another. Today they are either fully or partially resolved. But the focus of my three stories is not on resolution, arriving at an answer; rather I am concerned with process, the nature of conflict and how it is conditioned by the culture of each discipline.

I have a good friend and colleague who was diagnosed with lung cancer last year. Charlie is fit; he watches his weight, loves the out-of-doors, and tells me he has never touched a cigarette in his life. Despite his good behavior, maybe he once breathed in too much asbestos or perhaps he lived too near a hazardous chemical plant and that's what caused his malady. Surgeons have already removed one of his lungs and irradiated and chemoed what's left of the other. Poor Charlie seems to be a victim of bad luck—the result of another roll of the dice in a game of life scripted by Mother Nature's crapshoot. Was the inevitable clicking on of a genetic switch predestined to deal him this fate? We simply don't know. But there's a good luck side to Charlie's coin of fate. At this writing the chemo is working and he is in remission. Eventually of course, one will ask, why did Charlie die? Why did Tony die? Why did whoever is reading this book die? Isn't *why* the ultimate question about death and dying?

If luck does have something to do with it, then the whole world lucked out in the winter of 2004. As *Time* put it, "For one night it looked as if a killer asteroid was about to hit—but the Earth's number isn't up quite yet."[5] Two months earlier, astronomers tracking the asteroid 2004 AS1 noticed that the 100-foot diameter object was hurtling earthward on a collision course. A direct hit would be capable of generating the equivalent of a one-megaton H-bomb explosion! Early calculations (not made public for obvious reasons) had put the odds of a strike at one in four. Two days before impending impact astronomers had another look and they revised their estimates. On February 16, the space-borne rock passed within 8 million miles of our abode (around 33 times the distance of the moon)—a close call by cosmic standards, but not a record.[6]

Earth had dodged another bullet. I say another because this was not the world's first Chicken Little alert. It was only the most recent in a long chain of asteroid collision warnings that *Time*, CNN, and *USA Today* have been posting with great regularity for the past two decades. I own a set of asteroid actuarial tables put out by NASA that gives the probability of collision with a range of object sizes and the megatonnage of ensuing destruction that would result. For example, a 50-yard-sized object packs the equivalent of about 12 million tons of TNT. What are the chances of getting bombed by such an object? Several thousand to one in any given year. Meteor Crater (4000 feet across) in northern Arizona was created some 50,000 years ago by just such an object. A mile-wide asteroid collision would constitute a major disaster. If such an object hit central Europe, it would wipe out much of what lives on that continent. Fortunately, such an event happens only about once or twice in a million years. But, astronomers reckon, it is when we get into the wider-than-a-mile range that asteroid impact can wreak permanent havoc—degrade the world's climate, cause crops to fail, and render many species extinct. A five-mile object (we get whacked by one of those about every hundred million years) would easily be capable of producing mass extinctions.

My earliest recollection of the impending threat of a cosmic encounter goes back to childhood. I remember viewing the 1950s sci-fi film, *When Worlds Collide*, about the impending encounter of earth with Bellus, a passing star that had entered the solar system, and the reaction of a Cold War populace to the threat. Americans and Russians needed to cooperate to launch a crash space program to transport a significant sampling of *Homo sapiens* to a safer abode (Zara, the companion planet to the stellar interloper). One of the major issues was who would get to go. After seeing that movie, I recall lying awake for weeks worrying over the future of our planet.

Studying asteroids was once regarded as a boring, relatively uninteresting sideshow in the discipline of astronomy. Black holes, the expanding universe, and the exotic atmospheres and dazzling rings of giant planets were far more enthralling subjects. Every astronomy text had an obligatory "Solar System Miscellany" chapter where you could learn in a few pages all that was worth knowing about asteroids, the 40,000 odd fragments, ranging in size from dictionary-sized chunks to almost legitimate tiny planet status (up to 500 miles in diameter), that orbit the sun, mostly between Mars and Jupiter. Did this debris simply never coalesce to form a planet or had two or more once fully formed planets collided to create the asteroid belt? This question is still debated.

The asteroid fear factor didn't begin to materialize in the public conscience until the 1970s, when data from the unmanned interplanetary missions began to roll in. I'll never forget those first close-ups of Mercury and Mars—surfaces ravaged by impact from archaic storms of rocky leftovers from the early stages of formation of the solar system between 3.1 and 4.6 billion years ago. Later, Magellan's cloud-penetrating radar would prove that even an atmosphere one hundred times as dense as our own was insufficient to protect Venus's surface from blemishes acquired from space debris impact. The earth has a host of holes, too, like the Ungava Crater in northern Quebec, Canada (at two miles in diameter the largest directly attributable to extraterrestrial impact), and, of course, familiar Meteor Crater. Satellite technology has revealed more than a hundred others up to 120 miles across. One of the major results of the unmanned space program was the stark revelation that we live in a dangerous, violent solar system.

In the final decades of the twentieth century, asteroids have joined tornadoes, hurricanes, and earthquakes on the list of natural phenomena that we need to be concerned about when we hit the road for the office every morning. Some blame America's paranoia about impending disaster on millennial anxiety—fear of the great overturnings that are supposed to accompany the return of time's odometer to a string of zeroes. Watching Comet Shoemaker-Levy break up and plummet into Jupiter's atmosphere in 1995 only added to the angst. "When will it happen to us?" one viewer who tuned in to NPR's "Fresh Air" program queried of a panel of experts on which I sat during the dramatic evening of the live telecast of the impact.

It had happened before—big time! In 1979 physicist Luis Alvarez and his son Walter, a geologist, offered a stunning answer to the long-standing question, what killed the dinosaurs? They boldly announced that the sudden extinction of the dinosaurs at the boundary between sediments dating to the Cretaceous (K) and Tertiary (T) periods 65 million years ago was caused by an unlucky hit by an asteroid. Paleontologists had always regarded the demise of the dinosaurs as a mass extinction, because all the evidence they had gathered from the layering of fossils pointed to a sudden cessation at the K-T boundary—which means it happened in a relatively short period of time, geologically speaking (not much more than two million years).

Some bizarre hypotheses for dinosaur death had already been on the books: they died from allergic reactions to new plant forms or from diseases contracted through migration to different environments; they got so big they could no longer move around; mammals devoured every egg they laid; they got zapped by a sudden influx of cosmic rays; a worldwide increase in temperature caused testes in male dinosaurs to cease functioning and so they became sterile; and so on. But none of these largely speculative explanations really satisfied.

Along comes Luis Alvarez, who was anything but a major figure in geology. As a physicist, he had won a Nobel Prize for his work with University of California at Berkeley's cyclotron. He had also worked on radar systems in World War II and then on cosmic X-ray diffusion; he even investigated and analyzed the famous Zapruder film in the JFK assassination, opting for the single assassin scenario, and X-rayed Egyptian pyramids for the remains of ancient pharaohs. Like most physicists I know, he was on record as not feeling much excitement about geology; but he did little to dissuade young Walter from becoming a geologist. In 1980 the father acquired some interest in things terrestrial when his son brought to his attention a curious sample of sedimentary layering he had been studying in the north of Italy.

The impact theory would rest on the discovery of a thin layer of iridium found in the sample. Later it showed up worldwide in deposits lying right at the K-T boundary. Since iridium is not abundant in the earth's crust, the Alvarezes reasoned that the only ready source for it in the sediments would have to be extraterrestrial. They calculated that a large extraterrestrial body on the order of 50 miles in diameter caused this worldwide catastrophe. Initial impact must have been most unfortunate for anyone within a few thousand miles of searing landfall. Eventually a cloud of dust and vaporized meteoritic fragments created a nuclear winter as a foggy fallout of poisonous gas and acid rain followed firestorms and hurricanes. These events eventually drove down the earth's temperature enough to unleash a catastrophic effect on life forms that were especially environmentally sensitive.

Then came the smoking gun: a large crater, just about the size (180 miles) that would have resulted from a low-angle impact of a 50-mile flying object from the south, was identified in 1981 in the Caribbean just off the north Yucatán coast near the Mexican town of Chicxulub. In fact, the boundary of the impact crater is so clearly visible on satellite photographs (see Figure 10) that it's a wonder the feature had eluded discovery for so long. It yielded an impact candidate: shocked quartz and glassy spheres that fit the profile of impact ejecta were found in the vicinity. Dating the Chicxulub event has subsequently proven problematic, with one vocal minority placing the impact 300,000 years before the K-T boundary.

Radical new ideas in the scientific disciplines aren't often given a warm reception—especially when they are launched by outsiders. Take the theory of continental drift. Today it is as basic to the study of geology as Darwin's theory of evolution is to biology. Alfred Wegener, the meteorologist who proposed this theory in the 1920s, was branded a lunatic when he suggested that a breakup of Pangaea, once the only continent on earth, resulted in the creation of separate American and Eurasian subcontinents. Wegener died in 1930, but in the face of mounting evidence (similar rock formations found on coasts of continents that appear as if they could be joined together like the pieces of a jigsaw puzzle) his theory was resurrected in the late fifties and early sixties.

I recall being a member of the audience in 1963 when geophysicist J. Tuzo Wilson laid out the theory of continental drift in a public lecture at my college. I remember too that some members of our geology department in the audience heckled him. I was also present twenty years later when a member of the Alvarez team—an astrophysicist—in the same hall presented the evidence

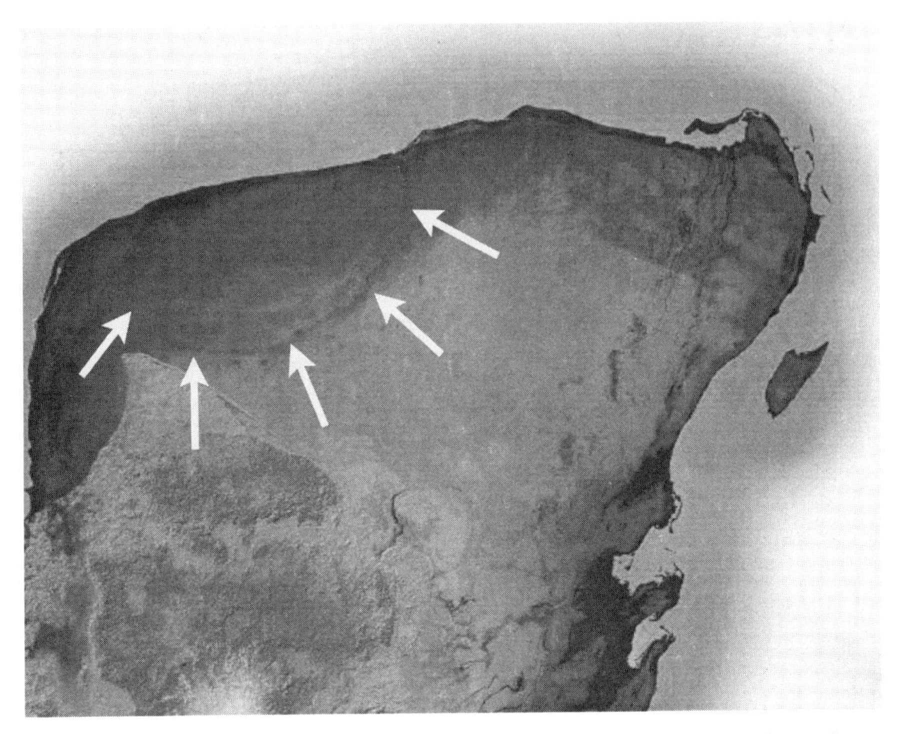

10. JPL satellite photo of the Chicxulub crater. The edge of the impact point, off the northwest coast of the Yucatán peninsula, is clearly visible as a quarter circle in the dark area at the upper left. (Courtesy, NASA/JPL–Caltech)

for what seemed at the time another especially radical idea. Seated next to a paleontologist colleague who silently listened to the entire exposé, I could almost swear that I visibly detected volutes of rage and anger smoke from his brain and rise skyward as he listened impatiently. After the lecture the paleontologist's fumes ignited into classroom rage as he spewed severely critical comments (based on evidence acquired from his own discipline) on the lecturer—who, he claimed, regardless of his knowledge of cosmic matters, was totally ignorant when it came to the subject of the demise of the dinosaurs. To posit an astronomical cause for a geological event was nothing less than an invasion of his disciplinary turf. To me, this was exciting!

Reaction to the Alvarezes' hypothesis in the journals was as swift and nasty as anything I had witnessed in the lecture hall. Here's a particularly vitriolic one:

> Environmental changes on this planet as recorded by the facies [rock faces] should be thoroughly explored before invoking the deus ex machina of strange happenings in outer space. . . . *It is intuitively more satisfying to seek causes from amongst those phenomena which are comparatively familiar to our experience* [my italics].[7]

One could add, Whose experience? Another castigating response goes even further, suggesting total fabrication, not to mention media mongering:

> Most of the "science" performed by the Alvarez camp has been so inexplicably weak, and the response to it so eagerly accepting by important segments of the scientific press . . . that some skeptics have wondered if the entire affair was not, on the impact side, some kind of scam.[8]

What everyone agrees on is that there was a relatively sudden climate change at the K-T boundary. Many—but not all—organisms became extinct. Ocean-bound mollusks and forams (single-celled forms with a shell-like covering) were hit hardest of all. Beneath all the name-calling, the thrust of many of the attacks centered around trying to show that the so-called K-T event happened at different times in different places and that the transition was anything but sudden. Furthermore, the iridium anomaly need not have been produced by impact. What alternative then? First, plenty of evidence points to increased volcanic activity around 65 million years ago, and these volcanoes could have belched up the mysterious element—an internal (geological) explanation rather than an external (astronomical) one. Second, an abundance of related continental drift was going on at that time. To paraphrase General Douglas MacArthur: Those old dinosaurs didn't die—they just faded away.

Further studies have since suggested that there were several great extinctions and that they may have been periodic—a 26 million year cycle has been proposed, perhaps induced by changes in the earth's orbit about the sun (the intrinsic point of view) or by the orbit of a "Nemesis" object, the sun's invisible twin, which lies far beyond the range of our telescopes, but regularly makes close passages through the cometary belt that surrounds the outer solar system, disturbs that material, and creates a flurry of impacts (the extrinsic alternative).

So where do we stand now? Too early to tell. Only twenty years have passed since the Alvarezes detonated their bomb. I know of no Roper poll among the disciplines, but based on my own rough sampling of a few dozen astronomer-physicists and geologist-paleontologists I've encountered, I would put the former group at 10:1 in favor of impact, with the latter group divided more evenly.[9] "Hard rock" geologists lie somewhere in the middle but tend to lean slightly "yea impact." (Some paleontologists openly declare they couldn't care less why the dinosaurs died. They are only concerned with why and how they lived.) Media coverage, by the way, goes 100:1 pro-impact and no wonder. In today's culture, catastrophe stories have legs. That's why tornado chasing logs more time than global warming on the Weather Channel.

Although the dinosaur controversy (like the quest for the Star of Bethlehem) isn't as socially relevant as the issue concerning the nature of the unborn fetus, it serves as an excellent example of disciplinary sword-crossing that takes place on the ideological battlefield of the sciences. Moreover, this sort of conflict between gradualists and catastrophists has happened before and in a variety of settings. When we look into these engagements we discover that along with a scientist's passion comes a scientist's bias, a bias built into a point of view conditioned by specialized training and bolstered by a gut feeling—call it faith—that one's disciplinary principles take precedence over those of another. This bias is also grounded in the experience that comes with a discipline's methodology and in the level of comfort acquired by familiarizing oneself with the workings of the natural world according to the laws and processes housed in that discipline. Recognizing our biases is important. We ought to do it more often.

My second story takes us back a century and a half to 1865. The American civil war has ended and a period of social reconstruction is about to begin on this side of the Atlantic. Over on the other side it is a time of immense advancement in the sciences as the Victorian era ushers in an attitude of great confidence that the exquisite design and order concealed in the womb of nature will soon be laid bare to the probing mind of the theoretician and the deft hand of the experimentalist. In England, Sir William Thomson, Lord Kelvin (1824–1907), a member of the top drawer of the hierarchical bureau of science, is at the peak of his game.

Kelvin was an accomplished mathematician, theorist, and engineer; discoverer of the basic principles underlying heat exchange (metrically-minded science still uses the temperature scale named in his honor); and creator of the Second Law of Thermodynamics, whence pop culture's much overused term *entropy*, or the inevitable natural trend toward disorder in the universe. White haired and bearded, always dapperly clad, the stately Scotsman cut an authoritarian persona at professional meetings as well as on the lecture circuit. Clearly, William Thomson commanded great respect among his peers; contemporaries said he was on a par with Sir Isaac Newton. As one fellow physicist wrote of him after attending a meeting of the British Association for the Advancement of Sciences:

> His personality was as remarkable as his scientific achievements; his genius and enthusiasm dominated any scientific discussion at which he was present. . . . He made the meeting swing from start to finish, stimulating and encouraging, as no one else did, the younger men who crowded to hear him. Never has science seen a more enthusiastic, stimulating, or indefatigable leader.[10]

Kelvin was about to play a pivotal role in one of science's great controversies. Like the dinosaur debate, it would pit the disciplines of traditional physical science, housed on its firm foundation of the universal application of natural law, against nascent geology, a field that was just beginning to walk erect, having only recently deployed its own set of quantitative tools and methodology. At issue, how old is the earth?

What civilization hasn't dealt with that fundamental question? Most would answer that the world had neither beginning nor end, or that it is as old as the most recent creation in a series of cyclic epochs. The start of the current epoch is usually arrived at by tabulating the many "begats" found in oral and written histories that trace human origins back to the ancestor gods. In the Greek *Theogony*, time begins when sexually promiscuous Gaia (Earth) and her son Ouranos (Sky), born out of his mother's affair with the underworld deity Erebus, lie together. Her labor is made difficult by Ouranos's jealousy. Zeus, their progeny, makes the descent of man possible by defeating the remnants of the forces of darkness, thus validating his legitimacy as ruler of the cosmos. In the Babylonian version of creation, *Enuma Elish*, a long line of descent follows Marduk's victory over negatively inclined forces of nature. Christian scholars pursue a similar creation path by tallying lineage spans acquired from a literal reading of Genesis. The most explicit version of such an origin came in Bishop Ussher of Ireland's 1664 pronouncement that the world was created on October 26, 4004 BC—at 9:00 AM. Despite the influence of science since then, more than half of all Americans still pretty much hold to the good bishop's account, as the poll I cited in the last chapter attests.

The practical-minded Greeks of the Ionian school offered still another perspective based on observation of nature's evidence. Xenophanes of Colophon (sixth century BC) thought the answer might be found in the sedimentary layers containing fossils of animals that once crawled the sea bottom—that is, the world is pretty old. When Heraclitus traveled to Egypt around 450 BC, he reasoned that the Nile River delta, which he so named because it was shaped like an upside down Greek letter *delta* (∇), must have taken a very long time to silt up.

To set the stage for the 1860s battle of the disciplines, we need to go back to the seventeenth century, when the world's first scientifically minded geologists, following the Ionian paradigm, began to make rough calculations of how long it would take mountains to erode away, rivers to cut their banks, and sediments to accumulate. Some concluded that these processes were so lengthy that there was neither a trace of a beginning nor a sign of an end; but by the mid-nineteenth century geologists were carefully measuring the thickness of individual strata of sediments, trying to tack a precise number onto creation. Compiling these measurements, together with careful observations of how long it took to deposit layers of a given composition and thickness, one investigator confidently concluded that the Ganges River Basin took 95,904,000 years to form. Of course such a calculation is fraught with assumptions. How do we know the rate of deposition of material doesn't change? What about floods, droughts, and the like? But when estimates from other kinds of calculations (e.g., the time it has taken the ocean to get so salty) arrived, and most of them pegged the age of our planet at around 100 million years, the new earth scientists felt they had acquired a sensible measure of security and satisfaction.

Meanwhile, totally unaware of these early geological studies, nineteenth-century physicists were concentrating on a few time-related problems of their own, among them the relationship between time and the orbits of the planets. If the earth were made of molten iron balls, one physicist reckoned, it would have taken but 75,000 years to cool down to its present temperature. Another calculated that if combustion alone were the source of the sun's light, it would burn out in just 1,000 years, which raises the question, how can you have an earth without a sun? Using the time-honored principle of gravitation, Kelvin picked up on an idea proposed by German physicist Heinrich von Helmholtz. Kelvin believed that the sun's power was generated mostly by the energy acquired when the particles that now make it up fell toward a common center—the way water tumbling over Niagara Falls gives up its energy to produce electrical power. His calculations pegged the length of time it took the sun to collapse down to its present size at approximately 25 million years. Thus spake the laws of physics.

Driving Kelvin's explanation was his own pristine Second Law of Thermodynamics. Simply stated, in any conversion of energy from one form to another, part of the energy will be converted into heat. Although the total amount of energy in any closed system (like the entire universe) will remain constant, in every physical process that results in a transformation of energy a small amount of energy will be lost to the outside. Entropy is a measure of the amount of energy available in a system for doing work (*entropy,* from the Greek *entropia,* or "turning into"). Stated in shorthand fashion: things run down. The Second Law of Thermodynamics, taken together with the First Law of Thermodynamics, which quantifies the relationship between mechanical energy and heat, provided Kelvin with a solid foundation for calculating the exact amount of heat available to the sun and the precise time scale for the dissipation of that heat. All Kelvin needed to do was to calculate, based on the observable evidence, all the possible sources of energy that could have gone into the formation of the earth and the sun—such as infalling meteors, chemical energy, and gravity—and then compute and sum up the rates of dissipation to figure out the earth's age.

In 1865 Kelvin issued the first direct challenge to uniformitarian geology by calculating that the earth's crust could not have had a constant temperature for the last several million years, because if it did, the loss of heat that we measure today from the surface, assumed to be constant over that period of time, yields an original temperature that would have been hot enough to melt down our entire planet.

For the geologists, these bizarre emanations from the universal laws of physics controverted the dominant uniformitarian philosophy that had held sway in their camp in the second half of the nineteenth century. The geological point of view reckoned that the earth is in a state of dynamic balance and that the forces that shape all geological processes are basically constant, if not for all time at least over very long periods of time—periods long enough to account for the slow evolution of the earth's visible features. To turn a Laplacian phrase (see Chapter 3), uniformitarians have no need of catastrophe to explain the existence of mountains, rivers, glaciers, and fossil beds.

Emphasizing the increasing disciplinary divide, theories emanating from the isolated disciplines of physics and geology had produced two radically conflicting earth time scales. One set a definite short-term limit, whereas the other was, for all practical purposes, temporally open-ended. At the outset all of this mattered little because physicists and geologists never talked to each other, at least not until a handful of them realized they were bumping up against one another on the same turf. Thank Kelvin's provocation for that.

With the hindsight of a century and a half of scientific discovery in hand, the paradox of a Grand Canyon older than the sun would become easily resolvable. It turns out that the sun's energy comes mostly from nuclear reactions, but that fact would not be known until the 1930s when accelerators enabled penetration of the atomic nucleus. Furthermore, hot iron balls—or hot anything else—isn't the only source of heat in the earth. Radioactivity contributes significantly—but that fact would not be understood until Madame Curie isolated radium at the beginning of the twentieth century. Radioactive dating of terrestrial rocks eventually would make it clear that the earth is 4.6 billion years old. That's plenty of time to permit the sun to expend only a tiny fraction of its available nuclear fuel supply.

Yet, these nineteenth-century geologists and physicists did not have access to this information—and for a long time were totally oblivious to results from anyone else's disciplinary camp. Therefore, they produced radically contradictory answers to the same fundamental question. With swords drawn, another battle of the disciplines was about to ensue.

On his own turf Kelvin's conclusions were widely accepted without question. After all, they were derived from time-honored universal physical laws that had been applied successfully over and over since the great scientific Renaissance. Many outsiders were equally impressed. After all, what authority in the halls of science could trump the universal laws of physics, the queen of the scientific disciplines? Even Mark Twain took time to weigh in on the controversy. He found it hard not to be influenced by the opinion of a great man:

> Some of the great scientists, carefully ciphering the evidences furnished by geology, have arrived at the conviction that our world is prodigiously old, and they may be right, but Lord Kelvin is not of their opinion. He takes the cautious, conservative view, in order to be on the safe side, and feels sure it is not as old as they think. As Lord Kelvin is the highest authority in science now living, I think we must yield to him and accept his view.[11]

Once you're endowed with a great reputation in one field, there's a temptation to sharpen your skills on alien turf. Actors become singers and singers try acting. Scientists are no different. Carl Sagan was confident enough to write a book about the brain, and Stephen J. Gould surprised many readers with his legendary insights into the world of Victorian opera and the game of baseball. A generation earlier, Britain's Astronomer Royal Sir Fred Hoyle (of Steady State fame) felt emboldened enough to make a foray into British prehistory by proposing that what the builders of 5,000-year-old Stonehenge really had in mind was the construction of a giant megalithic computer to predict eclipses. Nearly a century before that, Sir Norman Lockyer, discoverer of helium in the sun and editor of *Nature,* science's foremost journal, plied his quantitative skills to date Egyptian temples by matching the orientation of their axes with ever-changing stellar alignments, even as two prominent physicists—Sir William Crooke, a pioneer in vacuum physics, and Sir Oliver Lodge—conducted experiments aimed at proving that communication could be established between the spirits of the dead and the living.

Kelvin, too, was a reacher. "A great reform in geological speculation seems now to have become necessary," he announced as he began his lecture on Geological Time to a group of geologists convening in Glasgow in 1868.[12] The reception from the discipline he invaded was exactly what one might have expected. Kelvin was annoyed that the geologists, steeped in their silly idea of uniformitarianism, ignored his thermodynamic conclusions. In an encounter at a scientific meeting, a Scottish geologist told Kelvin to his face: "I am as incapable of estimating and understanding the reasons which you physicists have for limiting geologic time as you are incapable of understanding the geological reasons for our unlimited estimates." Kelvin's patronizing retort: "You can understand physicists' reasoning perfectly if you give your mind to it."[13]

Unimpressed with this scolding, another feisty geologist later wrote, "Let us . . . hear no more nonsense about the interference of mathematicians in matters in which they have no concern."[14]

Unimpressed by Kelvin's quantitative application of universal law, Thomas Huxley, an avid defender of the geological camp (and of Darwinism as well) responded with the words I have quoted in my second epigraph. Admittedly, it had taken a while for the geologists to lose their awe of the prowess of mathematicians and physicists and to develop their own quantitative methodology to a level of confidence and precision sufficient to challenge the theoreticians on firm ground. By the late 1860s earthbound scholars had discovered that there were many geological clocks and few of them ticked at the same rate. Furthermore, most geological processes were found to be cyclic—there is a rhythm in the earth. Mountains don't erode uniformly, nor are forests denuded at a constant rate. Long-term averages of geologic processes began to point to a much, much longer time scale than the simple application of physics and math would allow—500 times longer than what Kelvin had calculated—something on the order of 1.5 billion years.

Despite all the progress, the first serious interdisciplinary discussions and meetings on the earth's age didn't happen until 1915. Geologists, astronomers, physicists, paleontologists, and biologists assembled to present evidence and arguments from their respective disciplines, including new finds from the study of radioactivity and the energetics of stars, which pointed to some mysterious source of solar energy beyond gravity. Gradualism in the strictest sense was no longer tenable, but then neither was Kelvin's abbreviated time scale. As I noted earlier, radioactive dating did not put his idea to rest until the 1930s, seventy years after he unleashed it. But even then Kelvin remained unrepentant. Later in life he stubbornly told a colleague he thought his most valuable work lay in limiting the ages of the earth and sun.

Later critics offered more refined, direct attacks on Kelvin's total disregard of geological data. After tracing a long list of geologically derived estimates of the age of the earth based on its component parts, a 1916 textbook summed up the situation:

> The Age of the Earth—After the development of the contraction theory of the sun's heat, physicists, among whom Lord Kelvin was especially prominent, informed the geologists and biologists in rather arbitrary terms that the

> earth was not more than 25,000,000 years of age, and that all the great series of changes with which their sciences had made them familiar must have taken place within this time. But no one science or theory should be placed above all others, and other lines of evidence as to the age of the earth are entitled to a full hearing. If they should unmistakably agree that the earth is much more than 25,000,000 years of age, the inevitable conclusion would be that the contraction theory is not the whole truth. This is a matter of the greatest importance, for not only is it at the foundation of the interpretation of geological and biological evolution, but it bears vitally on the question of the age of the stars and on the past and the future of the sidereal universe.[15]

One historian of the age of the earth debate lay blame squarely on disciplinary habits:

> [T]he reluctance to give up accepted and hitherto productive concepts played important parts in the ensuing controversies.[16]

We may have worked out the problem of the age of the earth, but there are lessons to be learned here. Why was progress so long in coming? The geologists felt defensive because they were under attack and the physicists believed universal law (a powerful label when you think about it) overruled the uncertainties and errors that plagued the young discipline of geology. All of these attitudes were shaped by the professionalism and specialization that grew out of the nineteenth-century separation of the disciplines.

As we have seen, academic specialization divides the intellect in strange ways. And, as we will see in the final story in this chapter, sometimes the disciplines that conflict are about as remote from each other as one can imagine.

We begin on the Isle of Principe, Guinea, West Africa, May 29, 1919. One of the most famous predictions of Albert Einstein's general theory of relativity is that gravity bends light. If the light of a distant object, say a star, passes close to the edge of a nearby object, say the sun, while on its way to us, the sun's gravity should appear to slow down the passage of a star as it goes behind the disk and speed it up when the star reappears on the other side. The principle works about the same way as refraction in the earth's atmosphere, which has the effect of prolonging our sunsets by making the sun's disk seem higher above the horizon than it actually is. (Did you know that if the earth had no atmosphere, the sun you see perched on the horizon would actually lie just below it?)

Einstein made his prediction in 1916. Timing the appearance and disappearance of stars behind the sun would have constituted an excellent test of his theory. But the trouble is you can't see stars passing behind the sun except during a total eclipse. On May 29, 1919, nature provided a perfect opportunity for just such a test—a spectacular solar eclipse that tracked from Peru across the Atlantic through equatorial Africa, set to occur amidst a rich background of stars. Astronomers set up multiple stations along the eclipse path. The weather was glorious and the test was met perfectly. The stars would offer another test of Einstein's theory.

There had always been something strange about the faint companion star that trails Sirius, the Dog Star, the brightest star in the sky. Sirius B (alias "the Pup") was discovered in 1862 with an 18½-inch telescope, then the largest in the world. Astronomers suspected the mini-companion because they had observed a slight wobbling motion in the course of Sirius—as though it were being tugged from side to side by an invisible companion in orbit around it. But the Pup was an unusual Sirian attendant. Its dimness (it is 10,000 times fainter than Sirius) combined with its high temperature (close to 10,000°K, or about the same as its much larger companion) led to the startling conclusion that it must be very small (not much bigger than the earth) and very dense, about a million grams per cc—which means that one tablespoonful of Sirius B stuff would weigh about two tons!

Now if Sirius B were that dense, reasoned the relativists, anything coming close to it—including the light of its companion—ought to be profoundly affected by its strong gravitational field. Einstein's theory predicted that in this case gravitation's effect on light should show up in the spectrum of the star in the form of a very tiny shift of its spectral lines toward the red, or longer wavelength, end of the spectrum. In 1925, at a meeting of the International Astronomical Union in Cambridge, England, a dramatic telegram was read aloud confirming that the lines in the spectrum of Sirius B were indeed shifted—and by the exact amount predicted by the theory of general relativity. These events, widely reported in the post–World War I, news-starved tabloids, made an icon out of Einstein and a celestial celebrity out of Sirius. Anyone alive in

1925 knew about them. Incidentally, as is often the case with "spectacular" scientific discoveries reported in the media, it later turned out that astronomers hadn't really observed Sirius B's spectrum at all. Instead they were looking at the reflected light of Sirius A off Sirius B. The shifting of the lines? That was produced by variations in movement on the orbit—not by Einstein's theory about gravitational effect, which nonetheless has since been verified.

Fast forward a decade to Timbuktu in the early 1930s. Young French anthropologist Marcel Griaule is about to depart to points a hundred miles south to begin his fieldwork among the Dogon, a West African tribe then numbering between 100,000 and 200,000, who resided on the border between what today is Mali and Upper Volta.

The Dogon fascinated Griaule. They had a complex religion, extensive anatomical and physiological knowledge, and they were advanced agriculturally. The Dogon wove beautiful baskets and fabrics; they devised a system of written signs and numeration; and they created a sophisticated calendar and astronomy. All of this came as a revelation to the outside world, which then regarded sub-Saharan, "darkest Africa" as the stereotype of primitive savagery and cannibalism, certainly incapable of having produced any culture worthy of the term *civilized.* Griaule's informants, like Ogotemmeli, the venerable blind hunter and wise man of the village of Lower Ogol, seemed eager to share information about their lifeways. Griaule's work, much of it directed toward a popular audience, attracted lots of attention, particularly in Europe. The adventurous Frenchman came to be regarded as the champion of a proud African people who lived by a complex religion and cosmogony that made them the equal of any people.

The late 1920s and 1930s were also the heyday of Margaret Mead and the coming of age of the new discipline of anthropology, which was responsible for the discovery and appreciation of non-Western cultures and the idea that the Western world could profit, both morally and intellectually, from studying them. Anthropology as we know it had begun as an offshoot of natural history in Napoleonic France about 1800, when men of leisure organized nature walks: "Promenades in the country . . . made with the purpose of giving to young people an idea of the happiness which can result for man from the study of himself and from the contemplation of nature."[17] By the 1880s the study of culture had become a subsection of societies organized at the national level in Europe and

America and it entered the university shortly thereafter. America's first department of anthropology was installed at Columbia University in 1901 by Franz Boas, a renegade physicist whose way to the study of culture had begun with an interest in the perception of light and color among natives of Baffin Island.

It was not until the 1970s that a star-crossed Dogon myth caused the pathways of astronomy and anthropology to intersect. With his co-investigator Germaine Dieterlen, Griaule had published a report titled "A Sudanese System Concerning Sirius" in a prominent French anthropological journal.[18] Told to them after their twenty-year residence in the land of the Dogon by four respected elder informants, the story revealed the intimate secrets of Dogon cosmology, the center of which turned out to be, oddly enough, the bright star Sirius. The informants told the anthropologists that Sirius was really two stars. The smaller star, like the tiniest grain the tiniest thing in the sky (as they put it), orbits the bigger one in fifty years. By way of explanation one of them drew a diagram on the ground with a stick—a diagram that looks startlingly like an elliptical orbit (Figure 11). They named the star *pò tolo,* or digitaria, after one of their grains. The informants went on to tell Griaule and Dieterlen that the earth and the planets move around the sun—in elliptical orbits—that Jupiter has four moons, and Saturn has rings. The anthropologists reported this Dogon knowledge without the bat of an eyelash. And why shouldn't they? Immersed in their cultural studies, neither was aware of nor seemed to care much about how such knowledge could be gleaned from celestial observations made without a telescope.

Sometimes it takes a popular work to fan a spark of controversy into full combustion. Robert K.G. Temple's book finally caught the attention of astronomers two decades after Griaule and Dieterlin had published their results on Dogon astronomy. Temple had a novel explanation for the Dogon-Sirius enigma, which he revealed in his 1976 best seller, *The Sirius Mystery.*[19] His subtitle is a grabber: "Was the Earth Visited by Intelligent Beings from a Planet in the System of the Star Sirius?" Temple, who called himself an East Asianist and a linguist, stated that when he learned in 1967 that the Dogon were in possession of information about Sirius that was so incredible, he felt compelled to research the material. Several years later he concluded (he says reluctantly) that the information was over 5,000 years old, and that it was passed on to the Dogon from ancient Egyptians, who had been contacted by aliens.

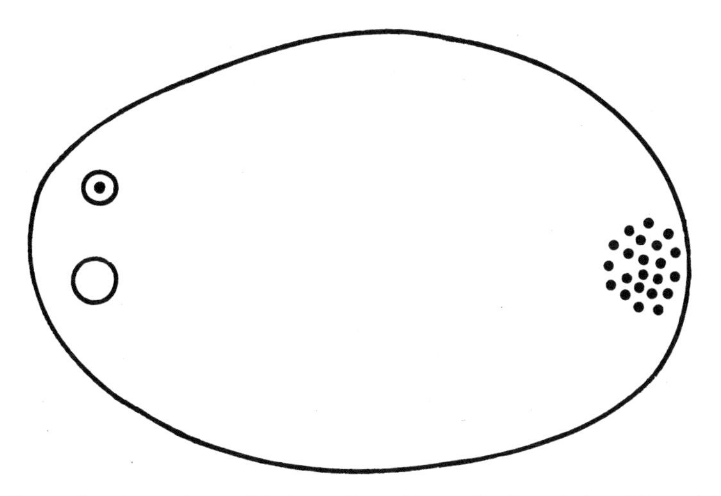

11. The Dogon drawing in the sand. Is it a stellar orbit or a fertile seed plot? (Kenneth Brecher, "Sirius Enigmas," in Astronomy of the Ancients, *ed. Kenneth Brecher and Michael Feirtag, 110)*

The evidence was there for all to see. Take the hieroglyphs. Osiris (or Sirius B substituting for Orion), companion to Isis (Sirius), is represented by an eye-glyph. The Bozo people of Mali call Sirius "the eye star," so there's one connection. Or the linguistics: the Egyptian word for dog (*auau*) sounds like a dog barking, as does the Sumerian word *bau.* The Egyptian god Anubis was dog-headed. Bau is the name of a Sumerian dog goddess. She was a sister of the fifty great gods, the Anunaki. The orbital period of Sirius B is fifty years, so there's another connection and so on. Dogon petroglyphs even picture visitors, the Nommo people, masters of the water, an amphibious race from Sirius come to earth in an ark (their spaceship) to give us the gift of water along with spiritual guidance. "The Nommo divided his body among men to feed them; that is why it is also said that as the universe 'had drunk of his body' the Nommo also made men drink." He also gave his life principles to human beings. "He was crucified . . ."[20]; "There will be a resurrection of the Nommo."[21] (Does this story sound familiar?)

When astronomers realized what the Dogon had allegedly accomplished they could scarcely blink. How could a people with little technology see an object several times fainter than the best human eyes could detect under optimum conditions? Furthermore, even if Sirius B were bright enough to be seen, it would be lost in the glare of its parent star. And what about the knowledge of Newtonian physics required to compute elliptical orbits? Where did they get *that* information?

The most logical explanation put forth by several astronomers was cultural transfer. Some Jesuit priest who had been reading the 1920s news sheets got into Dogonland before Griaule. As astrophysicist Ken Brecher described one scenario: the priest asks the Dogon about sky myths. After the informant tells him a few star stories, the priest replies, pointing his hand southward: "'Do you see that star? . . . [I]t is actually two stars, and the invisible star is the heaviest thing there is.' The Dogon promptly incorporated this information into their sky myths. Later, when the two anthropologists are told the secrets of the Dogon, all they get is a cross-cultural translation."[22]

This explanation extols neither anthropological method nor the integrity of indigenous religion and mythology. Would a Christian be so quick to alter symbols of the supreme savior if some outsider were to casually argue that his/her cross of crucifixion represents a monstrous evil force? But astronomers know little and care less about anthropological method, much less the meaning of mythic knowledge.

Akin to the explanation of cultural transfer is confabulation attributed to forceful questioning. Griaule is so keen to redeem Africa by demonstrating the complexity of its religious systems that he politely forces his informants to create new myths. Remember the school nursery scandals of the 1980s? Kids fantasized tales about being abused by their caregivers only after they were barraged with suggestively phrased questions.

I have actually witnessed confabulation resulting from my own work. In the 1970s we were studying an ancient round tower, the Caracol, believed to have functioned as an observatory and located at the Maya ruins of Chichén Itzá in Yucatán. In an article published in the 1950s, a famous archaeologist, Sir J. Eric S. Thompson, had commented on the Caracol's odd architectural plan by comparing the structure to a wedding cake perched on the square carton in which it came. Amused by this analogy, when I published my own report in *Science Magazine* in 1975, I chose that statement, which the author had also used to characterize other peculiar-looking, tasteless buildings, such as New York's Grant's Tomb and London's Albert Hall, as an epigraph. I left a copy of my report with the innkeeper at a small place near the ruins where I and my students stayed while doing the investigation. When one of my students returned to visit the Caracol several years later, she encountered one of the groundskeepers, a local, elderly Maya man who was sweeping the steps. Out of the blue she asked him, "Do you know why your ancestors built this building?" He replied without hesitation, "They used it for weddings."

But some anthropologists believed Griaule's method was flawed. The few who had lived with and studied the Dogon had found no evidence of a double star, a dense star, or a fifty-year orbit. So either Griaule passed on this news of Sirius B to the Dogon or he mistook other stars visible in the sky near Sirius for the invisible companion. After all, Griaule was no astronomer.

Novel explanations aside, there is no escaping the fact that Griaule's anthropological account attributes a knowledge of the Sirius system to the Dogon. But I get a sense that the discipline of anthropology wasn't given a fair hearing. My own opinion, shared with a handful of others who have taken the trouble to read up on Dogon mythology, is that the drawing in the sand and the whole Sirius myth has a lot more to do with sex than with astronomy or aliens.[23]

Astronomers don't care about sex—at least not in connection with the subject matter they study. Besides, connecting a distant star with an earthly grain seems a huge leap. After all, the stars are very far away. But for the Dogon the stars are as immediate a part of the environment as earthly things, so the equation *grain* = *star* might sensibly portray the one as a reflection of the other. A closer reading of Griaule's later work suggests that digitaria plays a pivotal role in Dogon mythology.[24] Digitaria isn't just a grain; it is a tiny, very dense grain, one of eight original grains—among them beans, rice, and several types of millet—given by God to the first Dogon ancestor. Digitaria excites men. Dogon legend has it that because it was too small and difficult to prepare, it is a forbidden grain connected with intimate feminine matters, such as menstruation. The legend, which faintly resonates with the Eve-and-the-apple story, says that it became the taboo grain, the one rejected by the first ancestor. But at a critical time, when all the other grains were exhausted, later ancestors breached the order by consuming it. For that reason they were deemed unclean and they exiled themselves from that region. Since then, digitaria came to symbolize the evil charms of a woman, which prevented her from conceiving: "[It] unwinds the evils of the good word round her womb and coils itself there in their place."[25] It is like menstrual blood—a sign of barrenness induced in the

12. A second diagram portrays the oval and its parts as female genitalia. (Marcel Griaule and Germaine Dieterlen, "Un Système Soudanais de Sirius," Journal de la Sociéte des Africanistes, *fig. 4)*

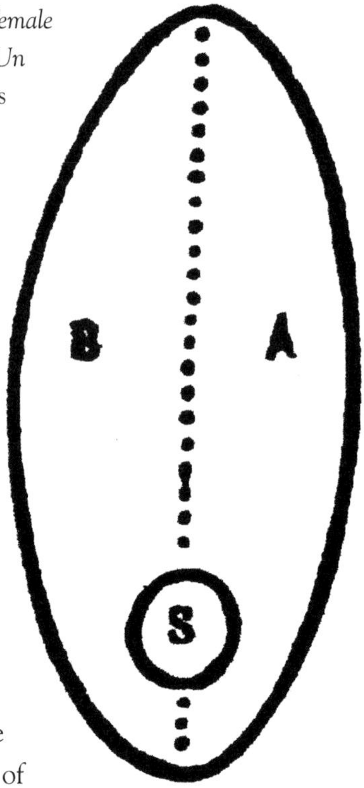

womb by these evil coils—and it is egg-shaped, not elliptical.

If you look at the drawing in the sand (Figure 11) closely, you'll see that there is nothing positioned on the "orbit." (Sirius and its dense companion are at the left, and the curious field of dots at the right have been taken to be Sirius's more distant third companion.) But maybe the drawing in the sand isn't an orbit after all. If you read Griaule and Dieterlen carefully, you'll also note that although they use the French word *trajectoire* ("trajectory") to describe this elliptical shape, they also call it the "egg of the world." And they state specifically that the trajectory symbolizes circumcision. In an oft-overlooked similar diagram taken from their notes (Figure 12), they even go so far as to identify the relevant anatomical parts. The oval with a line down the middle, for example, depicts female genitalia.

Twins are important in the Dogon myth. Sexual relations began with circumcision and excision, and *everything born has its twin*—an animal is born with every person, says informant Ogotemmeli. Every infant is born with twin sexes. In fact, digitaria is placed in the womb to engender the birth of twins. For the Dogon, twin births recall the ancient past, when all things came into existence in twos, thus symbolizing the balance between the human and the divine. Women who bear twins are touched by the ancestors and their children are forever the objects of the ancestor cult. Further linking digitaria with twins, Griaule had also identified the term *pò tolo* with the type of womb that produces twins. If at some point the Dogon had acquired information that Sirius had a companion, they might have been motivated to introduce it into their ritual because of its relation to twinness. Or they might have assigned Sirius a double simply because they knew that any important star must have one.[26]

If you think about fertility from an agricultural point of view, the diagram in the sand becomes a seed plot, and the line that bisects it (Figure 12) is the plowed furrow made by the farmer. The round object is the seed. As we have seen, sowing and reaping digitaria is redolent with sexual symbolism. The Dogon say that at the harvest the digitaria falls to the ground when it is beaten by the threshers, just as a woman gives her menstrual blood as a sacrifice due the earth.

Could there be a real environmental connection between the brightest star in the sky and Dogon agricultural affairs? I did a little checking and discovered that Sirius makes its first annual appearance in the eastern sky just before dawn about the tenth of July. This time is the peak of the rainy season in West Africa. And its last disappearance in the west happens around the first of January, the time of least rainfall. As Sirius passes around the sky from dawn to dusk, the rain waxes and wanes. The ancient Egyptians recognized that the timing of events in the environment could be scheduled by the whereabouts of Sirius. They noticed, for example, that the annual inundation of the Nile happened when the Dog Star made its first morning appearance. We still use the term "Dog Days" to describe the hottest days of the summer.

This information raises the question, Is it possible that West African people learned about the importance of Sirius from the high culture of ancient Egypt? Assessing the African contribution to science was part of the Afrocentric movement that swept the country in the 1970s. Many African Americans who had heard about the Dogon-Sirius enigma and the various explanations offered by the astronomers—that a missionary, or perhaps Griaule himself, had introduced modern astronomical knowledge to the Dogon (or worse still, that the Dogon were naïve enough to blindly accept something told them by foreigners)—were bothered by interpretations that necessitated contact with scientifically literate Europeans. From their point of view the way white men were proposing to solve the Dogon mystery constituted an offense to the traditions of their ancestors.

One response from the Afrocentrists came in a 1983 reader titled *Blacks in Science* intended for teachers and designers of school curricula.[27] It was edited by Ivan van Sertima, author of the controversial and influential (it went through twenty-one printings) *They Came before Columbus*, which had espoused a deep African presence in America long before slaves were brought over. Van Sertima had credentials; he was a professor of literary criticism and anthropology at Rutgers and one of the undisputed leaders of a movement designed to boost African American culture.

The first three chapters of *Blacks in Science* boldly declared that the Dogon mystery was no mystery at all. In one of the chapters, van Sertima traced the Sirius tradition back to antiquity, claiming to find drawings akin to Dogon representations of Sirius in the Egyptian hieroglyphs. A second reading of the record suggested to van Sertima that since the ancient Egyptians and the Nubians to the south possessed glassmaking technology, they could have had telescopes; or that telescopes weren't even necessary because the Dogon were endowed with eyesight superior to Caucasians because of their excess quantities of melanin. They didn't need to be told about Sirius's companion, if they could see it for themselves. Some of these chapters were adopted in high school science curricula in U.S. city schools. This book created a backlash in the national science education community at large. Here were minority groups spreading scientific illiteracy among themselves. As one critic commented: "Minorities are already greatly underrepresented in science and engineering. Teaching them pseudoscience will result in making it much more difficult for these young people to pursue scientific curricula."[28]

Temple's account—that this extraordinary knowledge came from extraterrestrials—was even more bothersome to the promoters of African culture. Black critics took the position that these explanations were all convenient excuses to deny the possibility that indigenous African people could have been skilled enough to have acquired it on their own—in other words, arrogant racism. (Astrophysicist Ken Brecher's statement, my third epigraph, was prominently cited.) But one could argue that Africanist attempts to explain white man's science through indigenous concepts is equally racist. And while all of this was taking place, I remember thinking that these efforts to scientifically rationalize a rich and wondrous Dogon mythology by making it sound West European was a kind of unnecessary subservience to science itself.

Today the Dogon controversy has quieted down. So has the dispute about the age of the earth, with the place of the decimal point finally having been set in stone. In the battle over what caused the death of the dinosaurs, the cosmic catastrophists seem to have won the day but I wouldn't bet the last word has been heard on that front. In its view from afar, history offers us clarity. Each of my three stories highlighted a collision of disciplines, the featured players possessing diverse training and interests. There are enough similar case studies to fill an entire book. But although the feuds about the death of the dinosaurs, the age of the earth, and the Dogon-Sirius enigma seem ideologically remote from one another, a common lesson weaves them together. As long as we all remain steadfastly focused on our own point of view, blinders firmly in place, more such battles will surely be waged.

NOTES

1. "Miscasting the Dinosaur's Horoscope" *New York Times,* editorial (April 2, 1985), A26.

2. Thomas H. Huxley, *Discourses, Biological and Geological Essays* (New York: Appleton, 1909), 335–336, quoted in Joe Burchfield, *Lord Kelvin and the Age of the Earth* (New York: Science History Publications, 1975), 84.

3. Kenneth Brecher, "Sirius Enigmas," in *Astronomy of the Ancients,* ed. Kenneth Brecher and Michael Feirtag (Cambridge: MIT Press, 1979), 109.

4. Stefan Collini, "Introduction," in C. P. Snow, *The Two Cultures* (Cambridge: Cambridge University Press, 1964), lv.

5. Michael Lemonick, "Chicken Little Alert," *Time* (March 8, 2004).

6. That record was set in August 2004 when a thirty-foot chunk passed within 4,000 miles of target earth.

7. Anthony Hallam, "The End-Triassic Bivalve Extinction Event," *Palaeogeography, Palaeoclimatology, Palaeoecology* 35 (1981): 36; quoted in James Powell, *Night Comes to the Cretaceous: Dinosaur Extinction and the Transformation of Modern Geology* (New York: Freeman, 1998), 26.

8. Charles Officer and Jake Page, *The Great Dinosaur Extinction Controversy* (Reading, MA: Addison Wesley, 1996), viii; quoted in Powell, *Night Comes to the Cretaceous,* 216.

9. Stephen Jay Gould wrote a favorable opinion of asteroid impact in 1985, arguing optimistically that it promoted cross-disciplinary fertilization. Stephen Jay Gould, *The Flamingo's Smile* (New York: Norton, 1985), Ch. 28.

10. Thomas H. Huxley, *Discourses, Biological and Geological Essays* (New York: Appleton, 1909), 46.

11. Mark Twain, *Letters from the Earth,* quoted in ibid., ix.

12. David Lindley, *Degrees Kelvin: A Tale of Genius, Invention, and Tragedy* (Washington, DC: Joseph Henry Press, 2004), 171.

13. Ibid., 173.

14. Ibid., 176.

15. Forest Ray Moulton, *An Introduction to Astronomy*, Rev. ed. (New York: MacMillan, 1916), 360–361.

16. Ibid., 216. Kelvin, clearly the goat in my second story, was also on the wrong side of the great nineteenth-century debate about the best way to electrify the world. In 1885 he ardently denounced the alternating current as dangerous to the world; cf. Jill Jonnes, *Empires of Light: Edison, Tesla, Westinghouse, and the Race to Electrify the World* (New York: Random House, 2003).

17. Translated and quoted in George Stocking, *Race, Culture and Evolution; Essays in the History of Anthropology* (New York: Free Press, 1968), 16.

18. Marcel Griaule and Germaine Dieterlen, "Un Système Soudanais de Sirius," *Journal de la Sociéte des Africanistes* 20:1 (1950): 273–294.

19. Robert K.G. Temple, *The Sirius Mystery* (New York: St. Martin's, 1976).

20. Ibid., 216.

21. Ibid., 214.

22. Brecher, "Sirius Enigmas," 110.

23. Very few anthropologically based explanations of the Dogon account have appeared in print. A particularly good one can be found in a letter-to-the-editor section of an obscure journal (Peter Pesch and Roland Pesch, "The Dogon and the Sirius Mystery," *Observatory* 97 [1977]: 26–28).

24. Marcel Griaule, *Conversations with Ogotemmeli: An Introduction to Dogon Religious Ideas* (Oxford: Oxford University Press, 1965); cf. E. Van Beek, "Dogon Restudied: A Field Evaluation of the Work of Marcel Griaule," *Current Anthropology* 32:2 (1991): 139–167, for a criticism of this work.

25. Griaule, *Conversations with Ogotemmeli,* 151.

26. Pesch and Pesch, "The Dogon and the Sirius Mystery," 26.

27. Ivan van Sertima, ed. *Blacks in Science: Ancient and Modern* (New Brunswick: Transaction, 1983).

28. Bernard Ortiz de Montellano, "Multicultural Pseudoscience. Spreading Scientific Illiteracy Among Minorities—Part I," *Skeptical Inquirer* 16 (1991): 46–50.

Part Two

WHAT WE SHARE

Beneath the conflict and disagreement, beneath the differences that separate discipline from discipline and Western science from the rest, and beyond our contemporary tendency to hone a narrow point of view in search of a bottom line, I believe that a handful of concepts concerning the natural world are shared by all cultures, concepts that inform the ideologies that underlie how we interpret nature. In the next five chapters we will explore some of these shared sensibilities. As we do, we'll discover that perhaps we in the modern West are not so isolated from the rest of the world.

F I V E

WHAT'S IN A NUMBER?

As I looked, behold, a stormy wind came out of the north, and a great cloud, with brightness round about it, and fire flashing forth continually, and in the midst of the fire, as it were gleaming bronze. And from the midst of it came the likeness of four living creatures. And this was their appearance: they had the form of men, but each had four faces, and each of them had four wings. Their legs were straight, and the soles of their feet were like the sole of a calf's foot; and they sparkled like burnished bronze. Under their wings on their four sides they had human hands. And the four had their faces and their wings thus: their wings touched one another; they went every one straight forward, without turning as they went. As for the likeness of their faces; each had the face of a man in front; the four had the face of a lion on the right side, the four had the face of an ox on the left side, and the four had the face of an eagle at the back. Such were their faces.

—THE BOOK OF EZEKIEL[1]

This mode of thought [science] presupposes an intimate parallelism between the mathematically expressible regimes of the heavens and the biologically determined rhythms of life on earth.

—HISTORIAN OF RELIGION PAUL WHEATLEY[2]

What I mean by a scientific astronomy is then a mathematical description of celestial phenomena capable of yielding numerical predictions that can be tested against observations.

—HISTORIAN OF SCIENCE ASGER AABOE[3]

Think of the number four. Four what? For most of us, numbers specify the quantity of things: four balls equal a walk in baseball, four weeks make up a month, four quarts a gallon, four quarters a dollar. Without something to count or quantify, numbers don't mean anything, right?

Galileo, the seventeenth-century father of modern astronomy, had a particularly low opinion of those who thought numbers alone signified anything. In his *Dialog Concerning the Two Chief World Systems,* completed in 1629, he uses the wise man Salviati to make this point by putting down the aptly named Simplicio for following old superstitious ways:

> SALV: To tell you the truth, I do not feel impelled by all these reasons to grant any more than this: that whatever has a beginning, middle, and end may and ought not to be called perfect. I feel no compulsion to grant that the number three is a perfect number, nor that it has a faculty of conferring perfection upon its possessors. I do not even understand, let along believe, that with respect to legs, for example, the number three is more perfect than four or two; neither do I conceive the number four to be an imperfection in the elements, nor that they would be more perfect if they were three. . . .
>
> SIMP: It seems that you ridicule these reasons, and yet all of them are doctrines to the Pythagoreans, who attribute so much to numbers. You, who are a mathematician, and who believe many Pythagorean philosophical opinions, now seem to scorn their mysteries.
>
> SALV: That the Pythagoreans held the science of numbers in high esteem, and that Plato himself admired the human understanding and believed it to partake of divinity simply because it understood the nature of numbers, I know very well; nor am I far from being of the same opinion. But that these mysteries which caused Pythagoras and his sect to have such veneration for the science of numbers are the follies that abound in the sayings and writings of the vulgar, I do not believe at all.[4]

For Galileo, numbers are descriptive adjectives, like the color red. They have no concrete meaning until you know what they are tallying, just as letters of the alphabet are meaningless until you put them together to form words that describe what things look like or how they behave. But among many cultures of the world, numbers alone did and still do carry an importance of their own. If you carefully read through the Old Testament passage quoted in my first epigraph, you'll note that the number four, taken by itself, does seem to hold some special significance. If you asked a person in the marketplace in the city of Jerusalem 3,000 years ago to think of the number four, you might get a different reaction from that of Galileo. And don't forget, our culture still recognizes lucky seven and unlucky thirteen.

Had you queried an ancient Maya in the maize fields surrounding eighth century Copán, Honduras, about the number four, he/she would likely have conjured up an image of a female deity with large, U-shaped eyes and a spit curl hanging over her forehead. Stela D of Copán (Figure 13) stands erect in front of a stairway at the north end of a 300-meter-long plaza studded with a dozen similar monuments. This larger than human-size monolith is dedicated to rituals that were conducted at the juncture of a series of important time cycles. If you take a close look at my reproduction of it, you'll note that eight squared-off images carved in high relief confront the eye at the top of the monument. Each depicts a humanoid figure engaged in what looks like a wrestling match with an animal: a bird in block A2, a toad in A3, a tortoise in B1, an elephant-like creature in B2, a grotesque humanoid in B3, an amorphous mass (probably the pelt of a jaguar) in B4, and so on. Closer inspection reveals that the humanoid figures appear to be carrying the animals.

Each figure looks like a porter caught in the act of transporting an animal. The men in frames B1 and B4 employ tumplines, common devices still used today by Maya peasants for carrying a load of wood or a sack of citrus by tying one's backpack to a band that presses tightly about the forehead, leaving the arms free to swing or perform other tasks. The porter in B2 cradles his conveyance under his arm, bearing most of the weight on his left shoulder. The youthful carrier in B3 grips his load by an ultra-long, deformed left limb (you can see its talons protruding in front of the bearer's face), while the old man in A2 almost seems to be caressing his avian cargo.

The porters are numbers and the burdens they bear are packets of time. Number nine (B1) is distinguished by the markings on his youthful chin. He wears a jaguar claw for an earplug. His freight is the heavy load of baktuns, 144,000-day periods consisting of 20 x 20 x 360 days. Although the Maya mathematical system shares the advanced features of place value and the concept of zero with our modern Arabic notation, there are two primary differences: first, the Maya numbers were written vertically with place value increasing from bottom to top; and second, the base of the system was 20 instead of 10, making the higher places 20, 400, 8,000, 160,000, and so on, rather than 10, 100, 1,000, 10,000, and so forth. When counting time, however, like their

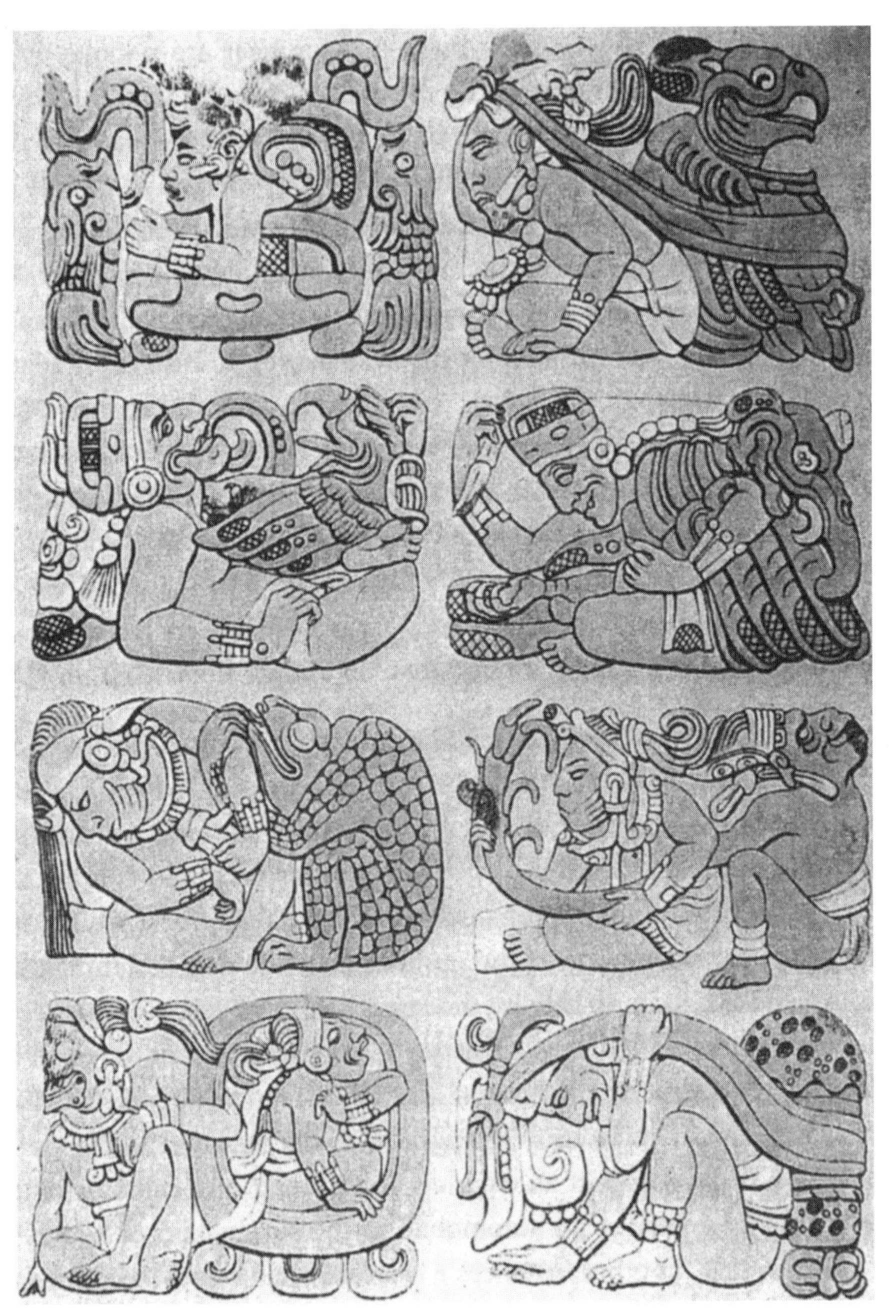

13. Stela D, Copán. Every Maya number was a lord or lady. Here each bears their own burden of time. (From Alfred P. Maudslay, Biologia Centrali-Americana, Archaeology *1 [1889–1902], pl. 48)*

counterparts in the ancient Middle East (from whom we acquired our base-6 method of counting hours, minutes, and seconds), the Maya developed a variant of this system. They altered the third place to 360 to achieve closer agreement with the seasonal year; the fourth place became 20 x 360, and so on. This vigesimal system presumably originated from the primitive habit of counting digits on both the hands and feet—not an unreasonable way to proceed among tropical cultures.

As in our number-word system, teens imitate single digits. The old god of number fifteen, shown in block A2, looks like the number five, which appears in frame B2, with the exception that the former's jaw is without flesh. Both wear a button-cap year sign in their turbans and both sport a single tooth. Number fifteen hauls katuns (scores of 360-day periods), while tuns, or 360-day years, are the burden of the number five. The pair of zero deities in frames A3 and B3, recognizable by the hand held over the jaw, tote *uinals* (or 20-day months) and *k'in* (or days), respectively.

Unlike its Western counterpart, the Maya zero represented completeness rather than emptiness. Temporally, it was regarded as the moment of conclusion of a cycle, just as we view the turning of a string of nines to zeroes on the odometer of an automobile as the completion of a unit that tallies distance. Often a seashell represents the Maya zero, perhaps because its roundness was intended to depict the closed, cyclic nature of time. The grasping hand, which bundles the days and years together into neat packages, also served as a zero. In most simple inscriptions, a dot was equivalent to one while a horizontal bar equaled five. All of these, including hand-on-jaw, probably descended from preliterate hand gestures: one through four consisted of the tips of the fingers while five was represented by the extended (completed) hand. In many cases numbers were highly stylized, the bars and dots adorned with bows, knots, crosshatched designs, bandings, and other frills.

For the Maya, numbers don't *describe* time—they are action figures who *carry* it. Fully transliterated, the numbered portion of Stela D reads: "it was after the completion of nine baktuns, fifteen katuns, five tuns, zero uinals, and zero k'in, reckoned since creation day, that . . . some important event took place. This interval, or bundle of time, which calculates out to 1,405,800 days, or approximately 3,849 of our seasonal years since the zero point in the Maya long count calendar (akin to a day in our Julian day calendar), was thought to have been delivered across eternity by deified numbers who served as the great bearers of time. We can think of Stela D as the resting place of the numbers at the end of their long journey. But only some will rest. Next year, number five will hand the baton to number six who will carry time through the next tun cycle; but number fifteen must persist for fifteen more tuns and number nine for four more rounds of katuns (they turn over at thirteen) before the odometer of time relieves each of them of their ponderous burdens. Such are the lives of the Maya numbers and, as you may have gathered from this brief glance at them on a Maya monument, we can go a long way toward understanding the common sense of another culture by studying its numbers.

Although the Maya personification of number and their use of a base-20 system seem strange and exotic, what fascinates me is that on opposite sides of an unconnected world, two systems that utilize both a zero and numeration by position should have even developed. In a way the Maya use of numbers for calculating vast quantities of time framed in a highly ordered cosmos, along with the idea that nature speaks to us through the language of mathematics, only fuels my appreciation of them—for these are concepts we share. The contemplation and uses of number show that they, and many other cultures of the world, are at base just like us. This resemblance is perhaps most obvious when we examine the main protagonist introduced at the start of this chapter: the number four.

Four orders the world. We see it in the sixth-century BC prophet Ezekiel's oracle that warns of the impending fall of Jerusalem to the exiled Israelites (my first epigraph). Remember the four horsemen of the apocalypse who appear to John in the Book of Revelation? The first rides a white horse symbolizing Christ, the second a red horse for war and bloodshed, the third a black horse symbolizing famine, and the fourth a pale horse signifying pestilence and death—all signs of the great cataclysm to come.

The Bible is full of symbolism, particularly number symbolism. There's 666, the number of the beast (but a scriptural fragment newly examined by reputable scholars suggests it was actually 616[5]); the opening of the seven seas; the twelve tribes of Israel each numbering 12,000, for a total of 144,000; and the non-utterable four-letter power word, the divine name of God: YHWH (Yahweh). Because everyone was afraid to speak it (wherever it appeared in the scriptures, it was usually read as *Adonai* or *Lord*), no one really knows how to pronounce it. *Jehovah* was the medieval pronunciation for "he exists" or "I am that I am." I'm convinced there's some meaning in the biblical number code that we haven't come close to understanding, probably because we don't take numbers very seriously. Whatever the code, the number four surely must play a prominent role.

There are four evangelists (and gospels) and four rivers in paradise. The four cardinal virtues (wisdom, bravery, temperance, and justice) stem from the pagan world of Greek philosophy. So do the four basic elements (earth, water, air, and fire), each in its own domain, that structure the material world that lies beneath heaven. Check out an old map and you'll have no trouble finding the four points of the compass and the four cardinal directions. You will also discover that four winds blow, one from each corner of the world.

Four organizes the sinister as well as the sacred. In magic, they call it the tetragrammaton, the name of the mystic number four, which was often a symbol representing God, whose name was expressed by the four letters of the power word. This was the quadripartite symbol par excellence for conjuring up any supernatural power, including the devil. That's because four is the number that describes the way in which all things, for good or for evil, are organized and constructed.

Four is the number of time as well as space. We recognize four seasons. After the fashion of the Roman calendar, we mark them at the solstices—the first days of summer (June 21) and winter (December 21) —and the two equinoxes—spring or vernal (March 20) and autumnal (September 22). The Celtic calendar, which has bequeathed us many of our contemporary holidays, pegged the four seasons at the midpoints of the seasons familiar to us: February 1 (whence our Groundhog Day and Candlemas between the beginnings of our winter and spring), May 1 (May Day between spring and summer), August 1 (the now defunct Lammas or first harvest between summer and fall), and November 1 (Halloween and All Saints' Day between fall and winter).

The Christian cross has four arms. Look down on a church from above—the way God sees it—and you'll discover that its architectural plan is informed by the cross, for the directions of the cross always intersect at the altar. Dig deeper and you'll find that most Christian symbols are borrowed from an earlier pagan era. The symbol of Christ's crucifixion derives from the Latin cross, which has its base stem longer than the other three. Before the Latin cross came the equal-armed Greek cross, which was preceded by the Egyptian cross, the Ankh, or symbol of life. It sported a loop at the top. The ancient Maya cross, which has the shape of the Greek letter *psi* (Ψ), is patterned after the tree of life with its branches arching upward. The swastika or gammadion, so called because it comprises four Greek capital-letter gammas that meet at a point, is a veiled form of the cross used in the early days of Christianity, when one's religious symbols were best kept secret. The swastika retained its underground status when it was adopted as a symbol among anti-Semitic secret societies in pre-Nazi Germany; the Nazis later co-opted it. The name *swastika* actually comes from a Sanskrit word (*svastikuh*) meaning "to be fortunate," "of good fortune," or "be well," in the sense of "Hail"—whence "Heil Hitler!"[6] Its design seems to suggest a whirling motion, the sun, a flame, a flash of light, perhaps even the movement of the Big Dipper around the Pole Star.[7] You can find the gammadion on tombs in Troy and at Etruscan ruins, on coins from Corinth, on carved Celtic stones, and in the art and architecture of China, India, and the Americas. (Figure 14 shows a few of these quadripartite symbols.)

Eastern religion also has a four-legged base. The Four Steps of Ascent to the Divine in Sufism—divinely inspired law, the narrow mystical path, truth, and divinely inspired intuitive knowledge—are paralleled in the Four Noble Truths of Buddhism: suffering characterizes all experiences; desire causes suffering; permanent freedom from suffering is possible; and the path of intellectual understanding and mediation taught by Buddha can permanently release us from suffering.

The quadripartite principle of organization goes well beyond Asia, Europe, and the ancient classical and Christian worlds. For example, the Aztec sun stone arranges space, time, and the theme of the ascendancy of Tenochtitlán, the Aztec capital, around the number four. The central visage is the face of the sun god, Tonatiuh. He is flanked by four calendar signs, each beginning with the number four and each pertaining to the times of great destruction that terminated the four ages that preceded the present one. Thus, in the first age a race of giants was eaten by ocelots; the second world ended in the winds of hurricanes; the third, in a fiery rain; and the fourth was done in by flood. The

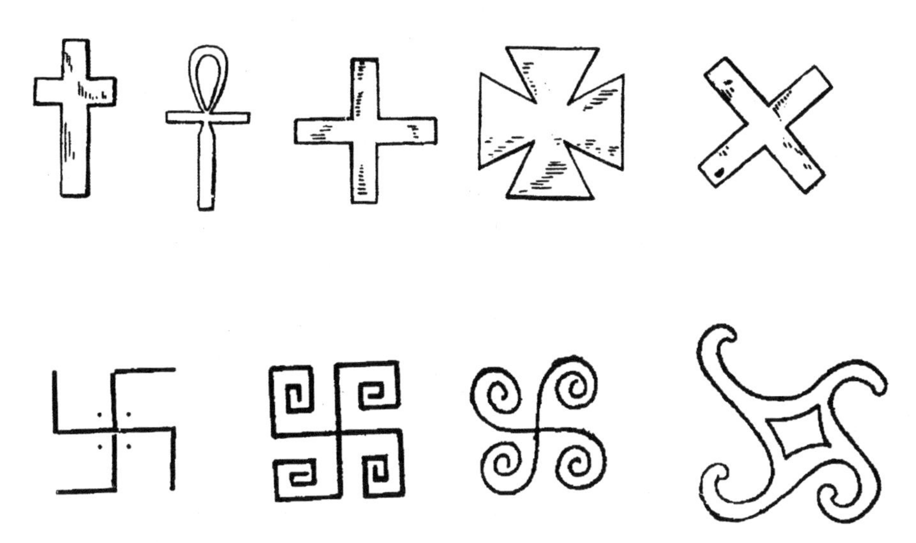

*14. A panoply of fourfold symbols; top row (*left to right*): Latin (Christian) cross, Ankh (Egyptian) cross, Greek cross, Maltese cross, St. Andrews cross; bottom row: forms of the swastika. (Samuel T. Wilson, "The Swastika,"* Annual Report of the Smithsonian Institution for 1894 *[Washington, DC: Government Printing Office, 1896], 769)*

capital city itself had a quadripartite plan, with four causeways connecting the Great Temple (see Chapter 6) in the central sacred precinct of the island city to the mainland. The Aztecs depicted this arrangement in the form of four streams of water that bound the tributary states on the periphery of the highland lake basin where the city is located to the center of power. This fourfold plan also resembles the way central Mexican picture documents, written before the arrival of Cortes, depict the universe (see Figure 27).

In the Mesoamerican world view the four sides of the universe are color coded. The eastern side is outlined in red, the color of the dawn; west, where the death god swallows the sun, is tinted blue, the color associated with the afterworld. A yellow sky, the color of maize, lights up the north; and green, the color of water and precious turquoise, demarcates the south. Color-coded cardinal directions also work in China, where they are associated with the elements, animals, and the seasons. East stands for blue-green, wood, dragon, and spring; west for white, metal, tiger, and autumn; north for black, water, tortoise, and winter; and south for red, fire, bird, and summer.

In his *Works and Days,* ninth-century BC poet Hesiod also tells of four ages that preceded the present one. First came the Golden Race, when all mortals lived in a paradise of abundance so that there was no need to develop agriculture or domesticate animals. Then came the Silver Race, a haughty, somewhat lazy civilization—at least too haughty and lazy to worship the gods. Next came the warlike Bronze age, populated by misshapen individuals with powerful secondary arms growing out of their shoulders. They were followed by the Divine Race of the Heroes, the demigods who immediately preceded us. But the Divine Race was wracked by horrible warfare. Quite contrary to our own, the ancient Greek version of history is a four-chapter, downward-spiraling, regressive slide into the present. We live in the aftermath—the Iron Race—as a people destined "never to cease from labor and woe by day and never to be free from anguish at night," for we must tend not only to our own needs but also to those of the gods who died on our behalf in the previous age.[8]

The Hindu calendar recycles creation in almost exactly the same way. The four ages, or Yugas, pass in a similar downward slide, with the added mathematical detail that they contract in a 4:3:2:1 pattern of duration, beginning with the 4,000-year-long Golden Age and ending with the 1,000-year-long Iron Age. The same is true in the world view of the Inca of the South American high Andes. Sixteenth-century chronicler Felipe Guaman Poma de Ayala depicts the four great epochs of change pictorially, the most recent signaled by the appearance of the Spaniards in their capital city of Cusco.

Archaeological evidence confirms that the central organizing principle in the Inca city was quadripartite. The highland capital of Cusco was centered about the sacred meeting place of two rivers. On that spot the founders built their Temple of the Ancestors, the so-called Golden Enclosure, or *Coricancha* in the native Quechua language. From this temple, four roadways lead to the four directions of the world. Here's how another chronicler, Bernabe Cobo, describes it:

> From the Temple of the Sun, as from the center there went out certain lines, that the Indians call *ceques,* and they made four parts conforming to the royal roads (*suyus*) that went out from Cusco.[9]

Cusco, by the way, is a later name. The Inca likely called their capital Tahuantinsuyu, which means the meeting place of the four *suyus.*

Although its original four-part divisions are scarcely detectable today (Cusco was almost totally demolished within a few years of the Spanish conquest), they are clearly visible in the pristine provincial city of ancient Huanuco, located in a remote area high in the Andes to the west. Archaeologists who have recently excavated there have brought to light a large, open rectangular space at the center, cut on each corner by a diagonal *suyu* that proceeds out beyond the boundary of the city. Huanuco's shape fits the chronicler's fourfold description like a glove. Many contemporary Andean people still organize their villages this way. They say that both the water and the people meet at the *crucero,* the crossing point in the center of every village. In the contemporary city of Misminay the four roads that extend outward converge with a road in the sky said to be accessible where the *suyus* meet the four extremities of the Milky Way at the horizon.

In Eastern thought the mandala (Sanskrit for "the plan of the universe") is the main symbolic diagram used in the performance of sacred Hindu and Buddhist rites and meditations. Practitioners say that the mandala is the central receptacle, the collecting point of all the universe's forces, a place where the mind may be guided through the processes of dissolution and reformation of the universe and the soul or self. There are many kinds of mandalas but they usually comprise a concentric circle surrounding a square with apertures that lead to the four directions. Each of the mandala's four sides has its own transcendent set of properties—a material element, a deity, a sacred animal, and a sense or human emotion. These powers become integrated in the circle's center, or the *axis mundi* (the axis of the world), the great mountain on which the sky rests. In the world of the practitioner, the mountain is symbolized by the temple, and mandalas are usually drawn on the ground before temple construction commences. Their dimensions, divisions, and overall geometry figure into the large-scale structure of the temple. For example, the Cambodian architectural masterpiece Angkor Wat's proportions are based on a grid system that follows three distinct numerical components of the architectural mandala: 12, 28, and 32—all of which are divisible by four.

Early twentieth-century psychologist Karl Jung saw properties in the mandala that transcended the Eastern wisdom that gave rise to it. He called it the "archetype of the true self," for the mandala unites the inner world of the psyche with the outer world of matter. The circle symbolizes wholeness and totality; the square, the unification of opposites and the relationship between the center and all that lies outside of it. By meditating on the mandala, he believed, you can bring your ego, the lowest level of your individuality, into harmony with the higher part of your intellect, the part that lies submerged in your consciousness. Historian of religion Annmarie Schimmel speculates that Jung may have been attracted to the four-part order symbolized by the mandala as an antidote to the chaos in German society during the Hitler era.[10] By giving this number a psychological function, as one historian of religion remarked, he did for the number four what Freud did for sex.[11]

What the mandala is to Hinduism and Buddhism, the sand painting is to the Navajo—that is, a pathway to spiritual knowledge. Almost always organized around four-component, four-color symbolism, sand paintings represent a detailed map of the cosmos. They are executed on the floor of the dwelling, or hogan (*hooghan*), to heal the afflicted resident by attracting one of the supernatural beings depicted in the painting thereby directing that being's healing power to the afflicted person seated upon it. In the Navajo world view, illness is thought to be a disruption of cosmic order and the only way to cure it is to reintroduce the afflicted one back into the cosmos. When the sick person is placed on the sand painting, he/she is healed as the power of the cosmos enters from the four directions and flows inside of him/her:

> A sand painting assists healing in four ways: it attracts supernaturals and their healing power; it identifies the patient with their healing power; it absorbs sickness from, and imparts immunity to, the patient seated on it; and it creates a ritual reality in which the patient and supernatural interact dramatically.[12]

The hogan itself has four support posts, cardinally arranged, and four areas of designated activity within.

As with the Navajo, the four-ness of the real world intrudes on us, even in our eminently practical age where numbers have been reduced to the mundane task of counting things. Dissatisfied with the three dimensions of length, width, and height, twentieth-century theorists have seen fit to switch to a higher reality consisting of four dimensions. That time can be treated on an equal footing with the dimensions of space wasn't obvious until the mathematical treatment of space-time appeared in Einstein's special theory of relativity. We pay a price by not being able to visualize such a universe, but the economy of the number four continues to satisfy us.

The list of examples of four-part symbols, ideas, and concepts may not be endless, but it is far longer than I've indicated. Why, in so many instances across the millennia and in such diverse cultures, do symbolic correspondences testify to a belief that the order of the world is based on the principle of cardinal directions and the number four?

My fascination with numbers began when I first learned how to write them down. I distinctly remember harboring a desire to personify them—tall, lean Mr. One; rotund Mrs. Two; angular-faced Mr. Four with his pointy nose. I imagined that each number was a self, with its own character. This is initially how I came to know them. When I grew up and read a little history, I realized what a throwback I was. I was reinventing the wheel. Pythagoras, the victim of Galileo's critical barbs, thought much the same twenty-five centuries before me, except he went much further. The more I learn about the Pythagoreans, archetypes of the Greek fascination with numbers as a way of describing nature (see Chapter 2), the more I am convinced that they were people who realized that playing with numbers is just plain fun. They were so taken with them that they single-mindedly sought to reduce everything they saw in nature to pure number; that is, to find the numbers that resided in the things that made up the material world rather than viewing numbers only as a convenient way to describe them the way we do.

Number, Pythagoras said, is the root of all harmony. He was the most famous among several early discoverers that harmonic chords have to do with the relative lengths of the plucked strings or stopped tubes that sound them. Take a string of any length and twang it. If you pluck a string twice as long you get the octave of the sound of the first string. Pluck a string 1.5 times as long (ratio 3:2) and you get a fifth; 1.33 times as long (ratio 4:3) sounds a fourth.

Seeking to extend this harmonic principle of order based on pure number to the universe at large, Pythagoras wrote: "There is geometry in the humming of the strings. There is music in the spacing of the spheres."[13] In other words all the bodies in our planetary system revolve around us at distances given by mathematical proportions. The slower ones sound in the base register and the faster ones emit higher pitched tones—all in proportion to the sizes of their orbits. (Two thousand years later astronomer Johannes Kepler would pick up on this idea. After a lifetime of playing with planetary orbits he would derive his famous three laws of planetary motion that paved the way to Sir Isaac Newton's discovery of the universal law of gravitation.) Why can't we hear the music of the spheres? Aristotle explains that sound can be perceived only by contrast with silence and, since this sound has been with us all our lives, we don't recognize it. We are, he says, like the blacksmith who, out of force of habit, is unaware of the constant noise that goes on about him.

In the Pythagorean perspective the numbers one through ten—those we can count on the fingers—are the essential ones. All numbers counted beyond four are based on these digits. Add the first four (1 + 2 + 3 + 4) and you get to the last one. The first four are the origin of all numbers, said Pythagoras: one, the point; two, the line, arrived at by connecting two points; three, the triangle, the simplest two-dimensional figure; and four, the most elementary solid body, the tetrahedron. Stack them up in an equilateral triangle like this:

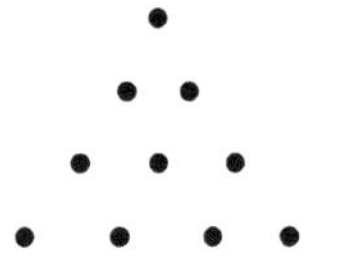

and you get the *tetractys,* a graphic expression of the $1 + 2 + 3 + 4 = 10$ idea.

Pythagoras also believed that the oddness and evenness of numbers reveal still more properties inherent in them. Compare the arrangements of the numbers on the face of a die:

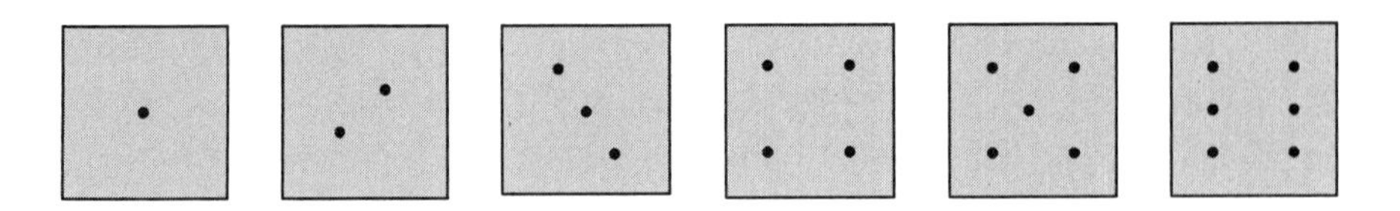

As prissy Plutarch moralized in the first century: the odd numbers have a "generative middle part" as opposed to the "receptive opening" of the even. The phallic 1, 3, 5, and so forth are therefore male numbers, and 2, 4, 6, and so forth, each with a space within itself as he put it, are female. In the chauvinistic world of Classical Greece this number sense meant that odd numbers equaled strength, good, and positive, but even numbers were weak, evil, and negative. And if that isn't enough of a slam on the feminine gender, the evil inherent in all women, also numerically derived, comes from the fact that you can reduce even numbers to the empty space and nothingness (read "primeval chaos") that preceded the present world by simply stripping away the outer points on the die.

Pythagoras's ideas were backed up by a religious, morally based philosophy. The pattern of all things to come was thought to be predetermined in the minds of the gods. Time, motion, heaven, and the stars all were created in the divine plan, which was based on pure number. As I hinted earlier, although modern science no longer embraces this philosophy, one of its legacies is the widely held belief that mathematics is the language through which nature speaks to us. (My last two epigraphs were written by contemporary scholars who have thought quite deeply about this matter.) So maybe number really *does* lie at the base of all natural law. And if you can discover the principles by which the numbers operate, you can place yourself in harmony with nature. Doesn't a person in physical and spiritual good health function like a well-tuned instrument?

Value-laden baggage aside, there are and were people in the world who believe that numbers have an existence of their own, independent from whatever transpires in the material world. But why does four hold such universal appeal? I believe there is something about the earth and the sky that connects us with that number. Try orienting yourself and you'll see what I mean. Orienting and orientation literally mean "easting," or facing in an eastward direction; the east side of the sky is where the sun comes up and consequently the direction sun worshippers faced. I am compelled to say *side* of the sky, because dawn doesn't always happen at the same point on the horizon. Sunrises oscillate seasonally. They sway like a pendulum, to the north of east (left) until June 21, our first day of summer, then to the south of east (right), where the sun touches land on our first day of winter, December 21, then back again.

We divide that swing into a two part northern half—spring and summer—and a complementary southern half—autumn and winter—which taken together add up to the four seasons. The angle of swing amounts to a sizeable chunk of the horizon—about 65° or 70° in mid-northern (or southern) latitudes depending on the horizon's elevation. Turn the other way and you'll notice that sunsets, along with the glow of dusk that accompanies them, likewise oscillate through a similar segment of the western horizon: to the extreme north of west (now on the right) on the first day of summer, then to the extreme south of west (left) on the first day of winter.

After you watch the sun go down, remain for the night show and you'll notice that the stars follow the sun downward in the west along oblique paths that cut the horizon (the angle is steeper the closer you get to the earth's equator). And if you reorient yourself by turning back to the east, you'll see that the stars rise along similarly inclined straight-angled paths. Anyone who bothers to look at the sky for an extended period (which excludes most of us) can't help but notice that there are two opposing, complementary sides to it. So there's a reason why people all over the world have developed sky-based dualisms, like day-light-life-east versus night-dark-death-west. These metaphors weren't simply dreamed up. They sprang from minds nourished by the senses and based on striking impulses gathered from the world around them.

Now, direct your eastward gaze beyond the limits of the swinging solar pendulum. Turn to the left (north), watch for a while, and you'll find that stars there neither rise nor set. Instead they pivot in perpetual motion about a fixed point located high in the sky (the farther north you travel on the globe, the higher the pivot point). Conveniently, that point is marked today by the moderately bright star Polaris, the pole star. If you point your left hand at that fixed point, drop your arm to the horizon and open a 90° angle from your extended left hand to the right along the horizon you'll land at the exact midpoint of the swinging solar pendulum—where the sun rises on the equinoxes, our first days of spring and autumn. Open the same 90° angle leftward and you'll strike the sunset point on the equinoxes. Not only have we learned where the four seasons come from; we also know why the fixity of the pole star served as the symbol of the eternal power of ancient Egyptian kings and Chinese emperors. That's why they carefully oriented their pyramids and temples that way. The ruling power on earth, like that in heaven, is ever present and eternal—the center of all cosmic activity.

The Egyptian pyramid directs the meeting point of its four sides skyward toward the center of its square base. It is a virtual mountain created by human hands, a representative of the cosmic mountain that swallows the sun nightly. Engineer Martin Isler argues persuasively that pyramids started out as simple mounds, their shape evolving from the primeval mound on which the sun god stood.[14] They became stepped, then smooth sided. Initially the Egyptians oriented them due east-west as a reminder of the link between creation and imperial rule made manifest in the daily movement of the sun; for like the king, the sun lies in the west only to be reincarnated as Horus, son of Ra, when he reappears in the east.

In China, the temple was oriented by similar principles. The emperor, as the heaven's son and successor, was an extension of celestial forces. With an eye centered on the fixity of the pole (the "golden nail" as they called it), rulers of the Jin dynasty imagined a heavenly Purple Palace made from four stars of our Little Dipper plus two others. This palace was surrounded by the Pei Chi, or "Four Supporters," each possessing a terrestrial social counterpart. The moon was governed by the crown prince, the sun was ruled by the great emperor, the planets were governed by the son of the imperial concubine, and the stars were under the charge of the empress. All the royal stars were placed near the celestial pole because, like the royal family's claim to rule, their presence in the sky was eternal.

Finally, turn to the south and you'll see low, sloping circular paths that behave opposite those of the north. If the focus of the north is up, then that of the south is decidedly down. Stars all seem to circle about an invisible point that lies under the horizon directly opposite the direction of the Pole Star. (In the southern hemisphere these north-south properties are reversed.)

The sky, then, really has four aspects, four sides arranged in paired opposites. These revelations, so obvious to the discerning eye, find expression in the very language we use to orient ourselves. For the Etruscans, the laying out of directions was not according to man but rather to the world itself. There was a front and a back to Etruscan ritual space, a left and a right in the earthly *templum* (whence our word for "temple"), the human-built representation of the celestial *templum* where the gods lived in the sky. "Four parts did it have: that toward the east *antica, postica* that toward the west; the northern part on the left, the southern on the right," wrote sixth-century sage Isidorus.[15] Clearly he was properly oriented when he wrote that statement. Varro, his second-century Roman predecessor, tells us precisely what the augur—literally, "one who divines by the observation of the course of birds in flight"—did when he laid out the earthly *templum*:

> When you look south for the seat of the gods, the east is on your left, the west is on your right. I believe it is for this reason that the augurs on the left are to be better than those on the right.[16]

Lucky right, with the corresponding unlucky or "sinister" left, is a Greek idea that emanated from four-directional symbolism. Curiously, the Etruscans seem to have had it the other way around.

In performing his rite, an augur first fixes a point on his left. Then he marks out a corresponding point on his right. In the middle, directly in front of him, he determines the sacred boundaries of the *templum* "by sight and by meaning."[17] Then he directs his gaze over the imagined city and the field that lies beyond it. He draws a line from east to west, then marks the limit in front of him as far as his eye can see.

Our way of laying out city blocks descends from the Romans, who acquired it from the Etruscans. When founding new cities or Roman colonies, Roman *gromatici* (surveyors) emphatically referred to upholding the ancient Etruscan practice of orienting their towns. The geographer Hyginus Gromaticus, for example, says:

> The *limits* were established not without consideration for the celestial system, since the *decumani* [the east-west streets] were laid out according to the sun and the *cardines* [the north-south streets] according to the celestial axis.[18]

Cardo, incidentally, is the Latin word for "hinge." In the mind of the Roman surveyor the north and south sides of the sky pivot about the east-west axis. His description of the basic coordinates of the grid plan continues:

> [They] divided the earth into two parts according to the course of the sun. The part situated to the north they called *right* and that situated to the south they considered *left* [clearly the viewer faces west]. The other line led from south to north and the parts on the far side they called *antica* and the parts on this side they called *postica.* And from these terms the boundaries of the temples also came to be described.[19]

The words for east and west in old Babylonian cuneiform translate as left and right, respectively, and the viewer's perspective seems to be from the north. To listen to the *Enuma Elish* (the Babylonian creation myth), the maker of the present world sounds like an architect. After conquering Tiamat, the source of disorder among the primeval waters, the hero Marduk uses the corpse of his defeated enemy to build the world. "Half of her he set in place and formed the sky as a roof. He fixed the crossbar and posted guards; he commanded them not to let her waters escape."[20] After determining the divisions of the year and establishing the stations of Enlil and Ea (the northern and southern halves of the sky), "[h]e opened the gates on both sides, and made strong locks to the left and right."[21] These are the mythical gates through which the sun rises and sets.

On the other side of the globe, the ancient language of the Maya tells us that their eyes beheld a cosmos that was not so different. The Yucatec word for north is *xaman,* or "on his right," and south is *nohol,* or "on his left." The difference here is that it is not the worshipper, but rather the deity, in the guise of the rising sun perched on the eastern horizon, who sets the directions.

The opening chapter of the *Popol Vuh, the Mayan Book of the Dawn of Life,* tells how the world was created by the female-male "mother-father of life." The act of creation is referred to as the emergence of "all the sky-earth." It comprised

> the fourfold siding, fourfold cornering,
> measuring, fourfold staking,
> halving the cord, stretching the cord
> in the sky, on the earth,
> the four sides, the four corners.[22]

Like heaven, earth, and our cities, our very bodies are also organized by the number four. Spread your legs and stretch out your arms, holding them just above the level of your head. If you are divinely proportioned, the tips of your fingers and toes touch the corners of a perfect square; the diagonals meet at your navel. At least, so said fifteenth-century mystic and philosopher, Cornelius Agrippa von Nettesheim. More legitimate characters (like Leonardo da Vinci) have also pointed out that the architecture of the body is like the architecture of the world. Robert Fludd (a contemporary of Agrippa) went further. Drawing on the old Pythagoreans, he likened the body to a stringed musical instrument, its pivotal points serving as steps that yielded harmonic musical chords equivalent to stages of the soul's descent from heaven into the body and its ascent back toward heaven.

Creators are architects. So what is it about the number four and architecture? Why do our buildings have four sides, our rooms four walls—not three or five? Surely our descent from the grid-minded Roman *gromatici* can't be all there is to it. The right angle has a kind of stability and symmetry, heaven-ordained or not, and it doesn't come across in any other kind of mensuration. True, round structures like the Taj Mahal and the Pantheon are elegant and symmetrical. As an astronomer, I've experienced the marvelous acoustics that reverberate under an observatory or planetarium dome; but round structures are a lot more complicated to build, especially on a large scale. Try making one out of Lincoln Logs or toothpicks and you'll appreciate some of the difficulties. (Stabilizing the curved arch is just one complication.) We continue the habit of building our universe the way we construct our buildings. And as we do so, we continue to discover the power and stability of the miraculous number representing the order of the material world—number four.

In his wide-ranging discussion of the Maya calendar, linguist-epigrapher Floyd Lounsbury speaks of two different motives regarding the Maya use of number. The better known preoccupation of the Maya's court arithmeticians and astronomers lay in using large numbers as a way of tying the lives of the rulers to their mythic ancestors of the past—the gods who created the world—and to anniversaries of creation events in the future, when the rulers themselves would ascend to the afterworld to become gods. But there is another type of numerology that involves smaller numbers, more like the ones we have been talking about, that cover brief intervals and time spans. The latter, Lounsbury thinks, is far more interesting, for here we discern numbers being chosen largely because of the nature of what they are—the very same kinds of choices the Pythagoreans made and Galileo discredited for being purely fanciful. We have every right to elevate Galileo to heroic status for helping to erect the foundations of modern science, an enterprise in which number plays a purely descriptive role in relation to physical reality. But all gains come with losses, and what we have lost in the bargain, I believe, is a sense of appreciation of the wonderful intricacy of equally valid and complex world views in which number, like word in the Old Testament, had a power all its own.

NOTES

1. Ezekiel 1:4–11.

2. Paul Wheatley, *Pivot of the Four Quarters* (Chicago: Aldine, 1971), 414.

3. Asger Aaboe, 1976 "Scientific Astronomy in Antiquity," in *The Place of Astronomy in the Ancient World: A Joint Symposium of the Royal Society and the British Academy,* ed. F. R. Hodson (London: Oxford University Press for the British Academy, 1974), A 276:23.

4. Stillman Drake, *Galileo: Dialog Concerning the Two Chief World Systems* (Salt Lake City: University of Utah Press, 1967 [1629]), 11.

5. www.worldnetdaily.com/news/article.asp?ARTICLE_ID=44169 (July 26, 2005).

6. S. T. Wilson, "The Swastika," *Annual Report of the Smithsonian Institution for 1894* (Washington, DC: Government Printing Office, 1896), 769. This lengthy, information-packed article (pp. 757–1011) is the most thorough treatment of this intriguing symbol that I have seen.

7. Zelia Nuttall, *The Fundamental Principles of Old and New World Civilizations* (Cambridge: Peabody Museum, 1900), 13–32.

8. Richard M. Frazer, ed., *The Poems of Hesiod (Works and Days)* (Norman: University of Oklahoma Press, 1983), 177.

9. Bernabe Cobo, *Historia del Nuevo Mundo,* Biblioteca de Autores Españoles, 91–92 (Madrid: Atlas, 1956 [1653]). For a translation see John Rowe, "An Account of the Shrines of Ancient Cuzco," *Ñawpa Pacha* 17 (1979): 27.

10. Annemarie Schimmel, *The Mystery of Numbers* (Oxford: Oxford University Press, 1993), 104.

11. Attributed to Victor White in ibid., 104.

12. Trudy Griffin-Pierce, "The Hooghan and the Stars," in *Earth and Sky: Visions of the Cosmos in Native American Folklore,* ed. Ray Williamson and Claire Farrer (Albuquerque: University of New Mexico Press, 1992), 115.

13. Arthur Beer, "Kepler's Astrology and Mysticism," in *Kepler: Four Hundred Years: Proceeding Held in Honour of Johannes Kepler,* A. and P. Beer (Oxford: Pergamon, 1975), 399.

14. Martin Isler, *Sticks, Stones, and Shadows: Building the Egyptian Pyramids* (Norman: University of Oklahoma Press, 2001).

15. Isidorus, *Etymologiae* 15:4, 7

16. Varro, *De Lingua Latina* 7:8

17. Ibid.

18. Carolus Thulin, ed., *Corpus Agrimensorum Romanorum Hyginus Gromaticus,* Biblioteca Scriptorum Graecorum et Romanum Teubneriana (Stuttgart: Teubner, 1971), 134 (emphasis added).

19. Ibid., 136.

20. Alexander Heidel, *The Babylonian Genesis: The Story of Creation* (Chicago: University of Chicago Press, 1942), 72.

21. Ibid., v, 9.

22. Dennis Tedlock, *Popol Vuh, the Definitive Edition of the Mayan Book of the Dawn of Life and the Glories of Gods and Kings* (New York: Simon and Schuster, 1985), 72.

S I X

TIME, THE CITY, AND A KING'S DILEMMA

To live in the city was simultaneously to have all cosmological knowledge presented to the senses, . . . [for the city] paralleled the cosmos in its layout, in its orientation, . . . in the way it transmitted the continuity between the body and the world.

—HISTORIAN PETER WILSON[1]

[T]he calendar of one city would in no wise [sic] resemble that of another, since the religion was not the same in both, and the festivals, as well as the gods, were different.

—NINETEENTH-CENTURY HISTORIAN
NUMA FUSTEL DE COULANGES[2]

I live in a house on the side of a hill at the edge of a small town in rural upstate New York. The center of town, about half a mile away by crow, is marked by the old Baptist church (a block south of the inn with the enigmatic painting discussed in the Introduction). When the town was founded around 1800, everyone worshipped there. The clock in the church tower still tolls out the hours to everyone within earshot. On quiet nights, that includes me (my bed lies next to an open window that overlooks the valley). Foggy nights muffle the clock's dozen midnight strokes, and the bell tones quaver liminally when a stiff wind blows. Sometimes when I lie awake I strain my senses to count them, wondering whether one, two, or perhaps three hours have passed since I put down my reading and decided, perhaps prematurely, to call it a day. In the daytime I often glimpse the clock's friendly Roman-numeral face and its pointy hands that seem to wave to me as I drive by the church on the way to the office.

If you stop and think about it, a clangorous mechanical device on a lofty perch—informing me as it does of my diurnal status on the longitudinal segment of the rotating earth that I share with fellow passengers in the Eastern Time Zone—is a strange concept. Even stranger is the idea that this benign, rather traditional clanging machine, designed deliberately to co-opt the beat of nature, was once a source of great social tension among the earliest urban dwellers.

My temporal whereabouts—and yours—emerged out of one of the great ironies connected with the development of mechanization in Western society: a religious, rather than a scientific, need was the motive behind the creation of one of Western society's paramount gadgets—the mechanical clock, which could precisely regulate the hours. Founded by Saint Benedict in the sixth century AD, the self-supporting Benedictine Order was one of the strictest when it came to discipline, which was achieved through ordering the lives of its members. The enclave strove for the perfection they believed came from living the most efficient life possible. Not a precious moment of our time on earth should be wasted. Inside the cloisters one activity followed another from early morning rise to bedtime: five hours of labor in the fields, punctuated by periods of writing and lots of praying. Everyone danced to the rhythmic pace of time marked by hourglass and sundial and announced by the signals of bells that tolled (and told) when to switch from one activity to the next. The rule of Saint Benedict specified when to "recite the hours": the *lauds*, the *prime*, the *terce*, the *sext*, the *none*, the *vespers*, and the *compline* in the waking hours, and the *vigils* and the *matins* at night. If we pray to God together, He will better hear our plea.

The mechanical clock was the final step toward the perfection of the good life here on earth. In the uncloistered world outside the monastery, it became an effective tool for organizing labor in the newfound market economy. The first mechanical clocks appeared in mid-thirteenth-century central Europe. These machines were an ideal response to a mercantile society's need to schedule the manufacturing and dispensation of goods and an efficient means for feudal lords to control the people who processed and distributed these goods.

Today nine out of ten Americans reside in an urban or suburban environment, but living in cities was quite novel then. The city changed the rhythm of human activity. As the machine age took hold, trade and commerce expanded. Workers migrated en masse from the country to get jobs. In the city they could become shoemakers, weavers, textile workers, or dyers; and they could bring home a pretty decent wage if they were well trained. But the strict schedule of the urban workday was a far cry from the rural peasant's former daily schedule, which had consisted of a list of chores that began with feeding the chickens and ended with bringing in the cows, all accomplished more or less in sequence and timed by the approximate rhythm of the sun in the sky. You started work at sun up and you quit when it was too dark to distinguish heads from tails on a coin. Although busy, the peasant's schedule was flexible. Clocks changed all that.

Cloche is the French word for bell. And had you lived and worked in Basel, Venice, Paris, or Naples seven centuries ago, the pitch, tone, and duration of a host of mechanically regulated bells would have audibly ordered your every activity. In the garment industry alone there were sorters bells, porters bells, cutters bells, dyers bells, and so on.

Historian Jacques LeGoff traces a major part of the development of mechanized labor time to the workers themselves.[3] Initially it was they who demanded longer working hours as a means of increasing their wages. LeGoff uses a 1315 ordinance from Arras, in the north of France near the present Belgian border, to illustrate the case. Assistants to fullers (clothes finishers who worked in textile mills) had complained that the increasing weight and size of the fabric they were required to handle was slowing productivity. A commission of delegates consisting of masters of the cloth trade represented them in one of organized labor's earliest wage disputes. The result was the establishment of a night shift. But employers complained that under the new labor regime, workers were cheating on their hours. This led to the installation of the first *Werkglocken,* a mechanically regulated clock with a very loud bell, which was placed in a workhouse in Ghent in 1324. The *Burgomeister* and factory employer levied severe penalties on anyone who tampered with the clangers, anything from 60 pounds for ringing the morning bell to call an assembly of fellow workers to the death penalty for using the bells to call a revolt.

The new version of the voice of time became so popular with management that within ten years mayors and aldermen were authorized to house the bell in the central church belfry of every northern European city with a textile mill. Clearly audible to the worker who lived in the suburbs, the clock, now entombing time in a wheeled escapement, told when to go to work in the morning, when to eat, when to return to work in the evening, and when to quit for the day.

Regardless of where laborers performed their tasks—whether in the vineyard or the weaving loom, at the shipyard or the mine, in the home or the shop—people came to resent the bells and to mistrust the people who rang them. This mistrust might have been deserved, given that the employer class also ran the town government. Time no longer seemed to belong to God; it was owned by those who presided over the city. The two-handed clock was even more angst-provoking than its one handed predecessor (Figure 15), which pointed only to the hour, and before that the silent sundial, with only a blurred boundary between the light and dark portions of a gnomon's shadow whispering the hour. (I wonder whether a hypothetical extraterrestrial alien could guess that the little hand on a wristwatch, like the dial on the plate of a sundial, represents the position of the sun east or west of the local meridian—another of Western civilization's many attempts to quantify the observation of nature.)

A robot child destined to control its makers, the clock in the tower at the town center became a symbol of conflict among the classes that made up medieval urban society. Bringing time indoors, putting it on the mantle, and literally taking time into our own hands by strapping it about our wrists, has only intensified the struggle over time. The lesson of the mechanical clock's history seems to be that if "time is money,"[4] surely whoever controls time wins the day.

Clocks exist in other cultures, too, provided we allow a less mechanical definition of *clock*. These "other" clocks grew out of the same process of transforming nature. Some complexly organized societies have even devised human clocks. For example, the Zuni *pekwin*, a shaman in charge of keeping the calendar, would perch in his shrine on a ridgetop high above his pueblo. There he carefully sighted the spot on the horizon where the sun rose and set on important days of the year, such as the midsummer green-corn festival. Likewise, the ancient Aztec version of the town crier "watched the stars at night in order to ascertain the hour," a Spanish chronicler of the sixteenth century tells us,

15. A mechanical clock with one hand (ca. 1450), predecessor to the Big Bens of today. (Ernest Edwardes, Weight Driven Chamber Clocks of the Middle Ages and Renaissance *[Altrincham, Eng.: John Sherratt and Son, 1965], 13)*

"that being his particular office."[5] The Nuer tribal chief of the southern Sudan raised his hand and pointed it skyward at the proper angle to indicate the time to eat or the time to bring in the cattle. (The gestural "hands" of our clocks are aptly named, aren't they?) Because these societies lived in close contact with nature, the cosmos is bound to be embedded in the way they organized their lives.

These examples of early time management demonstrate that the careful observation of nature lies at the foundation of many of our timekeeping systems, whether we measure the day, the month, or the year. For all of us, the whole business of making and marking time is a feedback process between people and the world around them. It begins when the eye makes contact with the most pristine and perfectly predictable part of the natural world—the sky. We take what we see and shape it to meet our human needs. We even make it a part of our built environment (as my first epigraph suggests), and that helps make nature a part of ourselves. But time's signals differ depending on where you acquire them. And what you do with time's information depends on the sort of place you want to make and what moves you to make it. As the French historian Fustel de Coulanges has written (my second epigraph): we don't all necessarily invent the same calendar.

The way the sky turns and what you see there basically depends on your geographic latitude. People near the equator see more stars than those who live in higher latitudes because they have visual access to both celestial polar regions. Tropical observers also note that the stars' pivots of motion (the celestial poles) lie close to the horizon. That means east-to-west celestial motion viewed from the tropics takes place along vertical paths. Stars rise and move straight up from the east side of the sky; then they cross the meridian (the north-south line that passes overhead) and plunge vertically downward on the west side. The path of the stars is different in higher latitudes, where the straight up-and-over-the-top action in tropical skies is transformed into a round-and-round kind of circuit. Viewed from latitudes far from the earth's equator, stars (as well as the sun, moon, and planets) rise and set along oblique tracks. In high latitudes, the pivot of all sky motion becomes the celestial pole, that remote point in the sky today attended by Polaris, which, we'll remember from the last chapter, always remains in a fixed position high above the north horizon (in the northern hemisphere)—30° up in ancient Memphis, 35° in Babylon, and 40° in Beijing. Now anyone who has been to the tropics and watched the night sky will understand why nineteenth-century skywatchers in the Gilbert Islands of Polynesia (near latitude 0°) connected the most important entity in their lives—the family household—with the sky above it. As they told informants, the sky was their house (we discussed their cosmological model in Chapter 2) and they used terms derived from their terrestrial houses to describe it. The horizon was *te tatanga,* "the roof plate." The sea was *maia-wa,* "the containing wall." Three *oka,* or "upright rafters," "supported" the eastern sky and three, the western. The *te taubuki,* or "main ridgepole," of the roof of voyaging, their name for the sky, was the north-south meridian. The sky really does mimic the structure of your house if you live in the tropics; but the domicile model just doesn't work for those of us who reside in the high latitude places.

What drives us: our heads or our bellies? One of the great debates among anthropologists focuses on whether centralized power and authority emanate from the desire to control important resources, or from the manipulation of ritual and ideology with less emphasis on material and political underpinnings. Most agree that the root cause (if we must insist on one) comes from a combination of the two. The question is, how much does each contribute? Let me demonstrate by using time as a resource. What if our January of 31 days were followed by a February of 28 days and a March of 25? In Java (now Indonesia), nineteenth-century Dutch colonists discovered a gnomon (there called a *bencet* or *banchet*) used by the natives to keep time. The natives used this gnomon to partition the year into twelve months of unequal duration, ranging from 23 to 43 days. This calendar is approximately what you'd get if you marked the passing months using equal segments of length traveled by the tip of the gnomon's shadow at noon as the measuring unit.

This variance occurs because in the tropics the noonday sun lies north of the overhead position for part of the year and south for the remainder. (Imagine starting your windowsill flower boxes on the south side of your house in spring, moving them to the north side of the house in late summer, then back to the south side in the fall.) In the latitude of central Java (7° south), when the sun stands on the meridian at noon farthest north of the zenith (on the

June solstice), the shadow length, measured to the south, turns out to be exactly double the length that occurs when the sun at noon lies farthest south of the zenith (the December solstice), when the shadow is projected to the north. Clever Javanese chronologists halved the shorter shadow segment and quartered the longer, thereby creating a twelve-month calendar. They started their year at the June solstice and counted two months until the sun passed across the zenith (at which time it cast no shadow). Then they counted four more months when the shadow lay on the other side of the gnomon. On the December solstice the shadow reversed itself, and so they counted four more months to the second zenith crossing. They completed their year with the two final months, when the shadow returned to the south side. Javanese timekeepers had seized upon a shadow-casting principle unique to their locality and they used it to set up the standard *bencet* used to unify the calendar all across the land.

I suppose you could say the Javanese were celestial determinists; their calendar was imposed on them by a unique set of heavenly circumstances that work only in their latitude, so they ran with it. But that isn't the whole story. The Javanese invention of the *bencet* to mark time was not just a cerebral recognition of sundial symmetry. An anthropologist working there in 1907 tells us that the seasons were principally designated by visible time markers in the cycle of the rice crop, which they cultivated on the hot plains by the sea. With a bit of forethought the descriptions of the husbandman's duties can be made to fit the schedule of the sun's movement in the sky as seen from central Java and reckoned by the *bencet*. What we have at work here is a schedule for organized labor that follows nature and is also structured by subsistence. For example, the farmer works especially hard at midyear (during the longer months) to facilitate the growth of his staple crop before it is left to mature in the later (shorter) segments of the year cycle. The development of the *bencet* is reminiscent of what happened in Europe in the mercantile age. The commencement of various work-related activities once was casually determined in Java, but the schedule later came to be designated metrically, precisely, and, if you will, technologically, the starting times being determined by the priests of various villages, each of whom consulted the local *bencet*. Unfortunately we don't know enough about the history of these people, but I will wager that installing the *bencet* in Indonesia created as much conflict and tension in the peasant community as did implanting the clock in the middle of the European city.

The human sense of time takes on a decidedly different quality when we pass from the household and agrarian village to the city. When you move to the city, you adopt big-scale architecture as the permanent living environment, which brings with it an acceptance of constraints that go along with other forms of expression that are more explicit, embodied, and objective than those you might have confronted in more open nomadic or rural societies. The urban calendar (and the cosmic viewpoint that comes with it) becomes the framework that serves the purpose of structuring, formalizing, and objectifying social life.

In most urban societies, the actual business of keeping time is, as in all specialties, left to the pros. In ours timekeeping is a burden shared by the U.S. Naval Observatory and the National Institute of Standards and Technology, who boast that their cesium atomic clocks are never more than a ten-millionth of a second out of whack. (A century ago the Naval Observatory set our clocks the old-fashioned way, by timing star crossings of the meridian.)

The ancient Aztecs were avid clock watchers. But unlike the mysterious Javanese chronologists, we know something about who they were, thanks to the Spanish priests who came over after Cortes invaded ancient Mexico City in 1519 to convert the natives. One chronicler tells us of a chronologist who was the king of a rival city of Tenochtítlan, the Aztec capital:

> They say he was a great astrologer and prided himself much in his knowledge of the motions of the celestial bodies; and being attached to this study, he caused inquiries to be made throughout the entire extent of his dominions, for all such persons as were at all conversant with it, whom he brought to his court, and imparted to them whatever he knew, and ascending by night on the terraced roof of his palace, he thence considered the stars, and disputed with them on all different questions connected with them.[6]

The Spanish historian goes on to describe the king's observatory. He tells us that it consisted of a walled enclosure situated on the roof of the palace and was just large enough for a person to lie down. There the astrologer would peep through (or over?) a device he describes as "a lance, upon which was hung a sphere." He continues, ". . . and I think that the reason for hanging a sphere of cotton or silk from the poles was for the sake of measuring more exactly the celestial motions."[7]

The astronomer in me would love to know precisely what this calendar keeper was up to. How was he taking the measure of time? As to what purpose was served (other than getting the exact time) by the data he collected, there is ample testimony in the opening sentence of the quotation. If Kepler indulged in astrology, why not an Aztec ruler?

Old Mexico City was once a city on an island in a vast lake (foolishly drained by the Spanish conquerors in the sixteenth century). Today, downtown Mexico City is marked by the *zocalo,* the principal marketplace. The *zocalo* is a vast (10,000 square yard) open square perpetually circumscribed by a traffic jam. The nation's Presidential Palace fronts it on the east; the slowly sinking, lopsided Cathedral of San Francisco, on the north. It is thick with vendors selling religious items, household gadgets, and mechanical toys. Clowns, mimes, and fire-eaters perform for both workers and tourists. The aroma of street food—barbacoa, churros, and tacos—and the sounds of street performers—mariachis, pop brass ensembles, and an occasional one-man band—fill the air. If you walk kitty-corner off the northeast corner of the *zocalo,* you'll discover a huge hole, two city blocks square, dug out by archaeologists. Within lie the ruins of the most important ancient Aztec's sacred structure, the Great Temple, or the Templo Mayor as the Mexicans call it. On this very spot, they say, the ancient founders of the city, a sturdy but impoverished tribe from the north, consecrated the space where they saw an eagle perched on a nopal cactus, a sign from their god of sun and war, Huitzilopochtli.

An old poem chanted by a patriotic eagle warrior (as all great military men called themselves) tells us that the Aztecs believed they were the triumphant and historically legitimate successors to all civilizations past, just as we believe we are the world's defenders of the old Greek ideal of democracy. The words to that poem, and many others like it, reveal their city's mandate: war and conquest.

With our darts,
With our shields,
the city lives.

There, where the darts are dyed,
where the shields are painted,
are the perfumed white flowers,
flowers of the heart.

The flowers of the Giver of Life
open their blossoms.
Their perfume is sought by the lords—
this is Tenochtítlan.[8]

Here was Aztec ideology at its most fervent, the ideology of a new state that had recently come to dominate its rivals in the Valley of Mexico. And it was steeped in cosmic imagery. Stone monuments strategically placed all around the city proclaimed the connection between militarism and time's message ordained in the cosmos. Take the cylindrically shaped six-foot-wide stone of King Tizoc. It depicts in the round the conquests of one of Moctezuma's predecessors. This person is shown as a man of action upon life's stage, sandwiched in between the subterranean world below and the sky above. For one who walks around it, the message is clear: We are gathering up the vital fluid of sacrificial blood from our captors to nurture our gods and to keep the heavens going round and round—for they are the source of our power.

Cityscape and skyscape join at the doorway of Aztec mythology. We enter that doorway when we begin to perceive the way the Aztecs made Tenochtítlan into a stage for acting out the purpose of their creation. Rounds of time, from the naming of the days of the short cycle of the twenty-day Aztec week to the long cycles of destruction and recreation, are framed in the most famous of all Aztec works of sculpture in the round—the great Sun Stone, described in the previous chapter. As the universe turns, the image of the flint-tongued sun god, Tonatiuh, at the center of the stone reminds all good citizens of the need to help keep the world in motion; that is to nourish the sun god with life's blood so that he may continue his celestial rounds, never to be extinguished by the threatening powers of eternal darkness. Around the figure of Tonatiuh lie reminders of the destruction that came to failed previous creations—destruction by fire, flood, and so on.

So goes the Aztec story of the founding of their capital city. But myth often has a way of getting embellished after the fact, retreaded to serve as a means of justifying the present by those in control. In the late fifteenth century, before Cortes arrived, Aztec Tenochtítlan had emerged as a powerful military state that ruled all the lesser kingdoms around the great lake from its central position. In the generation before the Spaniards' arrival, the Aztec state had undergone an explosive expansion, taking in tribute from as far away as Vera Cruz on the Gulf Coast to the east and from towns along the Pacific shore hundreds of miles to the south. Keeping it all together politically was becoming a problem. Some historians believe that the warlike aspect of Huitzilopochtli, the Aztec tutelary god, served as a means of cosmically validating Tenochtítlan's military posture—"we conduct war because the sky gods will it." Consequently, the Aztecs regarded him as the god of both the sun and military affairs. Huitzilopochtli's shrine shared the stage at the top of the Great Temple with the Temple of Tlaloc. Tlaloc was a deity associated with rain, agriculture, and fertility, who had been worshipped all over Mesoamerica for more than a thousand years prior to Aztec domination. This clever marriage of very different deities made manifest in architecture an alliance between the new powerful sun-war cult and the traditional agrarian-based deity.

When, after marching from the Gulf Coast in search of treasure, Cortes and his soldiers arrived on the scene in 1519, they were awed by the sight of the great city of more than 200,000 people. One of them was moved to write of this Venice of the New World:

> The great city has many wide and handsome streets; of these two or three are the principal streets, and all the others are formed half of hard earth like a brick pavement, and the other half of water, so that they can go out along the land or by water in the boats and canoes which are made of hollowed wood, and some are large enough to hold five persons. The inhabitants go abroad some by water in these boats and others by land, and they can talk to one another as they go. There are other principal streets in addition, entirely of water which can only be traversed by boats and canoes, . . . for without these boats they could neither go in nor out of their houses.[9]

The diarist speaks of the fine houses in the city that possessed large dwelling rooms and exquisite flower gardens and where the vassals of King Moctezuma and many rich citizens resided. He also tells of seventy-eight temples scattered throughout the city.

But when it came to the central sacred precinct of the gleaming city, Cortes himself was unrestrained in his awe for the pagan spectacle he witnessed there:

> Amongst these temples there is one, the principal one, whose great size and magnificence no human tongue could describe, for it is so large that within the precincts, which are surrounded by a very high wall, a town of some five hundred inhabitants could easily be built. All round inside this wall there are

> very elegant quarters with very large rooms and corridors where their priests live. There are as many as forty towers, all of which are so high that in the case of the largest there are fifty steps leading up to the main part of it; and the most important of these towers is higher than that of the cathedral of Seville. They are so well constructed in both their stone and woodwork that there can be none better in any place.[10]

Although the Spaniards were awed by the temple's splendor, they were horrified by the bloody scene of human sacrifice to the gods that took place at the front of the Great Temple:

> [T]here were sounded the dismal drum of Huichilobos [Huitzilopochtli] and many other shells and we all looked toward the lofty Pyramid . . . and saw [men] being carried by force up the steps, and they were taking them to be sacrificed. When they got them up to a small square in front of the oratory, where their accursed idols are kept, we saw them place plumes on the heads of many of them and with things like fans in their hands they forced them to dance before Huichilobos and after they had danced they immediately placed them on their backs on some rather narrow stones which had been prepared as places for sacrifice, and with some knives they sawed open their chests and drew out their palpitating hearts and offered them to the idols that were there, and they kicked the bodies down the steps, and the Indian butchers who were waiting below cut off the arms and feet and flayed the skin off their faces and prepared it afterward like glove leather with the beards on.[11]

Human sacrifice is scary—the gore associated with the spectacle of priests garbing themselves in human skin, the scale of it all! That's why we're so morbidly fascinated by the Aztecs, even if ultimately we dismiss them with a repulsive wave of the hand for committing such acts of savagery and barbarism. Little wonder the Spanish chroniclers portrayed the Great Temple as a house of demons (Figure 16). But if we look beyond the immediate act and probe Aztec myth more thoughtfully, we begin to understand that Aztec rituals of sacrifice have as much to do with their desire—much like our own—to seize the moment.

To an Aztec, to manipulate time would be synonymous with controlling the sun. They believed that the act of creation took place through human sacrifice. Aztec legend has it that time began in the ruined city of Teotihuácan

16. *The Aztec Great Temple is pictured as a house of demons in post-conquest documents. (from* The Florentine Codex, *trans. Charles E. Dibble and Arthur J.O. Anderson, Book 2, Part 12, fig. 884; reprinted by permission, ©1963 by the School of American Research, Santa Fe)*

(which means "city of the gods"), the place where the gods made the first sacrifice, thereby setting the sun in motion. Says a chronicler,

> This is plain: that there in Teotihuacan, they say, is the place; the time was when there still was darkness. There all the gods assembled and consulted among themselves who would bear upon his back the burden of rule, who would be the sun.[12]

But fueling the unborn sun would be no easy prospect. It required that one of the deities must throw himself into the fire and reemerge as the newborn sun. Nanauatzin—bravest of all—volunteered, and when the sun burst forth, like a bird hatching from an egg, he wavered from side to side and flickered on and off before setting himself on his eternal course.

> But when the sun came to appear, then all [the gods] died there. Through them the sun was made to revive. None remained who did not die (as hath been told). And thus the ancient ones thought it.[13]

All Mesoamerican people regard time as the image of the sun itself. Recall that the Maya word *k'in* means "sun," "day," and "time." We may view all sunrises and sunsets as mere demarcations between day and night; but to see night and day through Aztec eyes you need to imagine that every dawn symbolizes a recreation of the world, a whole new beginning. Aztec religion taught the people that they had a role to play. Because only *they* descended from the gods, they must help the sun to rise, for time could be reborn only through human action. Only humans could supply the blood of sacrifice necessary to keep the sun on his course. Placing offerings and sacrificing themselves and their subjects to the gods in the Great Temple was part of the ritual. Performing these rites was a hugely theatrical enterprise, as the chronicler's description attests. And it had to be done at the right time.

> Then to her Patron Saint a previous rite
> Resounded with deep swell and solemn close,
> Through unremitting vigils of the night,
> Till from his couch the wished-for Sun uprose.
>
> He rose, and straight—as by divine command—
> They, who had waited for that sign to trace
> Their work's foundation, gave with careful hand
> To the high altar its determined place.[14]

British poet William Wordsworth tells us that when he wrote this poem in 1823, he was watching stonemasons as they lay the foundation of Rydal Chapel in Westmoreland. We know that, like Rydal Chapel, many of the great cathedrals of Europe were oriented to the rising sun on the day of the town's patron saint, an orientation that often determined the layout of an entire town. Recall that the root word, *orient,* gives away the meaning of the concept: *orientation* means "deviation from east," a place so chosen because it represents sunrise on the day of equinox, where the sun was born to start the year (*east* translates as *ush* in Sanskrit, meaning "to burn").

Like their European clock-watching contemporaries, the Aztecs sought nothing less than to capture time and bring it to the center of their city; however, their mechanism would not be a clanging clock but rather a signal transmitted by light via an exquisite piece of sacred architecture.

I have often tried to imagine Tenochtítlan's architects and city planners grappling with the difficult problem of manipulating time, of capturing the precise moment to conduct an elaborate sacrificial ritual and delivering it to the proper place, the sacred space in front of the Great Temple where the people of Tenochtítlan could all assemble to witness the moment. On Sunday morning Christians ring bells; at the hours of Islamic prayer a voice from the minaret sings out; but for Moctezuma, king of the Aztecs, only the light of the sun—time itself—would be sufficient to serve as the silent message from the gods to commence the ceremonies.

An explicit statement buried deep in the Spanish chronicles offers a clue to what the king and his astronomers were looking for. I believe it also hints at the dilemma that ensued when they tried to keep time on its course in the Great Temple:

> On this day they flayed all the enemy prisoners, and dressed themselves in their skins; and they dedicated the festival to Tlatlauhqui Tezcatlipoca. . . . This festival fell when the sun was in the middle of [the temple of] Huitzilopochtli, which was the equinox, and because this was a little twisted, Moctezuma wished it torn down and straightened.[15]

I found that this important festival, called Tlacaxipeualistli, was celebrated at the end of the second month of the Aztec year, which translates to March 13. But that's the date reckoned in the calendar in use before the Gregorian reform, which wasn't instituted until 1582. When we add the customary ten-day correction to put the old calendar in synch with the present one, the date of the ritual of sacrifice is March 23—which is very close to the spring equinox. Evidently the chronicler had it right.

One fact seems clear regardless of whether the equinox date mentioned by the chronicler refers to the festival or to the sun's position: some serious structural modifications needed to be made on the building during the reign of King Moctezuma. (To complicate matters, there were two Moctezumas, but let's leave that problem aside for the moment.) Since it is tied to the equinox sunrise, the alignment mentioned by the chronicler would have had to be exactly true west-east, where the sun rises on that day. As the sun's path has changed imperceptibly since Aztec times, it should behave no different today than it did then. If you want to test whether the building really captured the sun's movement as the chronicler said it did, all you need to do is measure it.

Back in the early 1970s, when I learned from the archaeologists that a wall representing the seventh and last phase of the building had been exposed by excavations dating back fifty years, I was eager to have a look at it. The Aztecs had a habit of building over earlier structures every time a new ruler came to power—an architectural technique that neatly preserves history. So, armed with surveying equipment, I traveled to Mexico to make the first precise measurements on the Templo Mayor. Much to my surprise I learned that the Great Temple was not at all oriented due west-east as I had expected; instead its axis was skewed approximately 7.5° clockwise; that is, east of north by south of east. After pondering this result, it struck me, could this gross misalignment have something to do with Moctezuma's dilemma?

The answer, I thought, lay in the problem that the chronicler tells us faced the Aztec architects: to make the building function in a practical manner in timing the ritual, they would have needed to get the sun to register its equinox position as seen by an observer looking up from the plaza below. I calculated that on the equinox date the rising sun would have been viewed through the space between the twin temples from ground level clockwise when it reached an elevation of some 22° above the natural horizon—but, because the sun moves along a slanted path as it climbs up into the sky, only if the axis of the Templo Mayor deviated clockwise by 7.5°.

I had looked into what the chroniclers had said about the actual height of the building at the time of the conquest. Making such measurements today

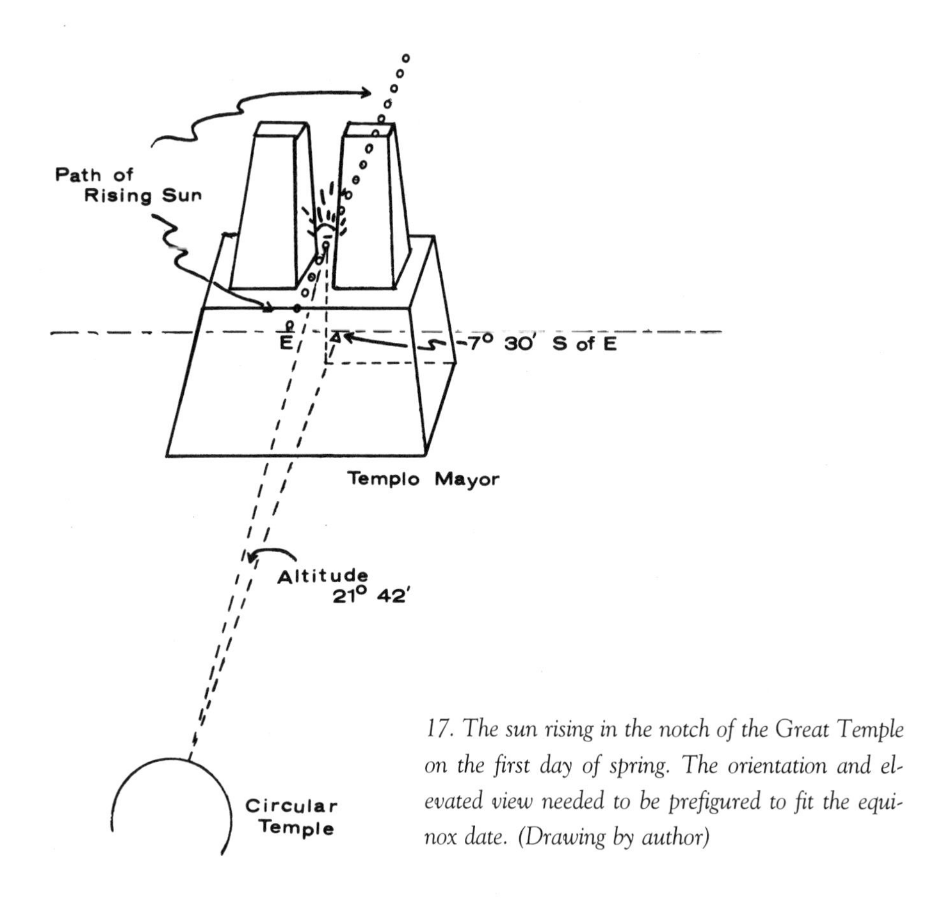

17. The sun rising in the notch of the Great Temple on the first day of spring. The orientation and elevated view needed to be prefigured to fit the equinox date. (Drawing by author)

would be useless because the structure has been sinking at the rate of several millimeters a year into the soft lake bed drained centuries ago. In fact, to get to the ruins today, tourists need to climb down a two-story flight of steps from street level. So a 22° elevation (as Figure 17 demonstrates) turns out to be fairly reasonable for an above-the-horizon solar appearance in the space between the temples on the first day of spring as seen from ground level. No further information was available until electricians working on the new Mexico City subway in the vicinity of the Templo Mayor in the late 1970s broke through some rubble and penetrated several earlier phases of the building. They also happened upon a huge carved circular monolith—a stone depicting the disembodied female lunar deity, Coyolxauhqui.

This accidental discovery, together with the pride of native blood that flows in the veins of more than three quarters of all Mexican citizens, led to the biggest archaeological dig in the history of Mexico—the excavation of two huge downtown city blocks with the goal of reconstructing the greatest of all temples erected by Mexico City's ancient ancestors. (Can you imagine city officials approving the excavation of all of Wall Street in New York, the Market District in San Francisco, or Michigan Avenue in Chicago for Native American artifacts?) As serendipity would have it, Mexico City's big dig would be supervised by the aptly named Eduardo Matos Moctezuma, likely a distant relative of the king on his mother's side. The indefatigable Matos and his team would burrow through no fewer than six building phases of the Great Temple, each arranged one atop another like multiple skins of an onion. Working with twenty-three other archaeologists over the next decade, the team would uncover, study, and register more than 100 offertory caches containing more than 10,000 tribute artifacts: delicately chipped obsidian blades; turquoise masks of Tlaloc, the rain god; alligator, jaguar and eagle bones; figurines; musical instruments; knives of flint; feathers; food; and textiles. Some precious trade items came from as far away as the ancient land of the Olmec on the Gulf Coast, the distant Maya of Yucatán, and, of course, the mother culture of Teotihuácan.

I followed the progress of Matos's architectural restoration with avid interest, wondering which way the early, inner layers of the sacrificial temple would be oriented. Would they be more or less twisted out of line than the outer Phase VII—graphic evidence of the king's dilemma? When these remarkable excavations and restorations were close to complete, at Matos's invitation I returned to the site in the mid- and late 1980s to take more measurements. The new data showed that the facades of all earlier phases of the temple that could be measured were oriented in almost exactly the same direction as Phase VII, the one we had measured a decade earlier. This was not what I had expected. Either the chronicler was wrong or the king's problem was not as simple as I had imagined.

Given the way the Aztecs reckoned time, would the sun continue to keep its annual appointment at the top of the building under these conditions? I needed to ask, if Aztec timekeepers measured their counting year of 365 days (by actually tallying days[16]), say, from the March equinox, and if they compared this time cycle with an orientation year of 365.2422 days (the time it takes for the sun, coming from the same direction, to return to the space between the twin temples), then how much disagreement would arise out of the two different methods over an extended period of time? My calculations revealed that it would have taken less than a decade for Aztec chronologists to recognize that the temple's alignment was slightly out of kilter with the sun, which would drift slowly *to the right* of its original equinox position at the top of the temple as the years passed.[17] If Huitzilopochtli were to be knocked off his pivot, the result could prove disastrous. Clearly there was a problem. Was this side step in the solar choreography the source of Moctezuma's situation? I found my answer as soon as I looked at Matos's latest plan of the lower interior layers of the temple.

The site plan (Figure 18) clearly showed that both the interior north wall of the Phase II Temple of Huitzlopochtli (the one on the right, or bottom in this figure) and the interior south wall of the Phase II Temple of Tlaloc (the one on the top left) were skewed markedly out of line. The skew lay in a clockwise direction and it amounted to about 4°. My calculations showed that a 4° shift (about eight times the size of the sun's disk) translates to approximately forty years' worth of accumulated error in real time. Clearly the king's clock needed adjustments. What could the king do to accommodate the rightward drifting motion of the sun? His court astronomers and chief architect would undoubtedly have been placed under enormous stress in grappling with the problem. At least for a while they could have bought some time by ordering a shift in the foresight, or viewpoint from the plaza below, toward the left. But the space between the twin temples is less than a yard wide, and if you tried to sight it from a point some sixty yards away down on the plaza, the sun would miss the notch completely about thirty years after the structures had been initially aligned to the equinox.

What other adjustments could the king have made to enable the sun to arrive at its proper place at the top of his temple? I think this is where the tearing-down procedure referred to by the chronicler may have come in. One way to repair the malfunctioning ritual clock would be to dismantle the original south wall of the left-hand temple, add a yard and a half to its east (rear) wall, remove a yard and a half from its west (front) wall, and then reconnect the two baselines by rebuilding the south wall. And that is exactly what we find in the skewed interior walls of Phase II.

Recall that two kings carried the name Moctezuma. If the skew was a Phase II problem, the chronicler is probably talking about a dilemma confronted by

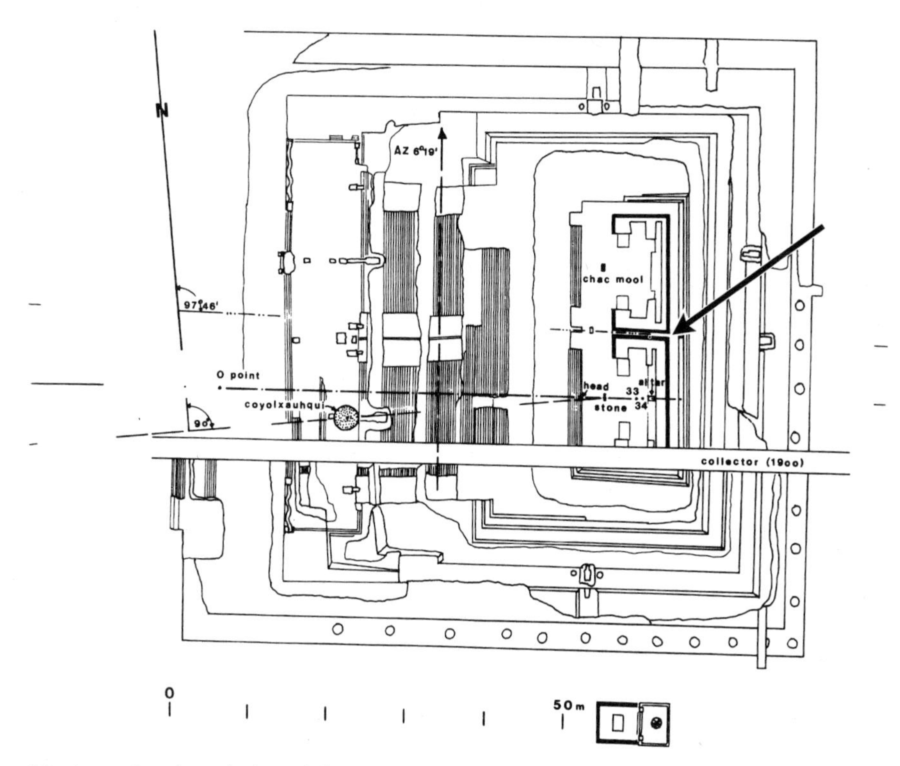

18. An archaeological plan of the Great Temple's early Phase II structure shows a skewed notch between the shrines on top (see arrow). *Was it a temporary solution to Moctezuma's dilemma? (Mexico, INAH)*

King Moctezuma I, or Moctezuma Ilhuicamina (ca. 1440–1467). Matos and his fieldworkers had dated the Phase II building to sometime before 1528 based on an Aztec date carved on an artifact cemented into that level of flooring. Since the orientation of the later phases of the temple stayed the same, in the years following Moctezuma I's reign no other changes took place. But my figures suggested that the drift rate of the sun would have provoked some sort of further architectural alteration shortly after the turn of the fifteenth century. Why were no adjustments made? I think there are at least two possible answers to this question. First, perhaps by the time of the reign of King Moctezuma II, or Moctezuma Xocoyotzin (1502–1520), the Aztecs no longer needed to keep time by sunwatching. We know that by then the Aztecs had written almanacs in their codices to keep track of time. But somehow I just cannot imagine the

Aztecs letting go of the sheer drama that comes from capturing time from the gods directly. There is a second possibility: maybe the later Moctezuma was indeed contemplating further adjustments to his great temple of sacrifice when he became preoccupied with other more pressing matters. (If you read the chronicler's quotation carefully, you'll see that he expresses King Moctezuma's *intentions* to tear down the temple and rebuild it.) The pressing matter was, of course, the Spanish invasion, which would destroy both Moctezuma and his temple. Could the off-schedule sun have been perceived as a sign of Moctezuma's inability to control it? Might it also have been a source of prophecies regarding Moctezuma's downfall?

The importance of keeping time in the Aztec city extended well beyond the sacred precinct. To create Aztec hegemony beyond the watery confines of their capital, the kings had built into the Templo Mayor a time axis that extended twenty-five miles from the ceremonial precinct at the center of the city to its periphery, where people lived who paid tribute to them. When I traced the alignment of the Templo Mayor beyond the plaza, across the lake, and beyond to the distant eastern horizon, I discovered that it passed directly over Tlalocán, the mountain home of Tlaloc, the deity worshipped in the other structure atop the Great Temple. The Aztecs believed Tlalocán (today called Cerro Tlaloc, elevation 12,500 feet) to be the source of all water, because during the rainy season the dark clouds of late spring afternoons first gathered over Tlaloc's mountain home. Remains of a shrine survive on the summit. There, chroniclers tell us, Aztec subjects offered young, weeping children as a sacrifice to pay their debt to the water god. The water-filled mountain of Tlaloc was part of an environmental clock that marked the place where the sun set during the twenty-day month that led up to the spring equinox, when the sun rose between the twin temples on top of the Templo Mayor—perfect timing for the Aztecs!

We tend to think of keeping time as a scientific pursuit, but so much of what I've learned about the ideology that underlies the Aztec practice of sacrifice tells me that their attempts to capture time in their city—to "seize the moment," as I like to put it—were conditioned by a multitude of factors both rational and humanistic. I believe the Aztec struggle with time and the orientation of the Great Temple was an accommodation between the old mountain-fertility cult of Tlaloc, via a visible link between Tlalocán on the mountain periphery and the Great Temple, and the new sun and war cult of Huitzilopochtli at the center of the sacred precinct. This link was sealed by attempts to lock time into the structure of the city via the movement of the sun. But I also see in Tenochtítlan another kind of accommodation—one between symbolic structures in a mythic landscape revealed in stories about the Aztec gods and another that involves locating and arranging their counterparts in the visible land and skyscape of a great city. For me it is this sort of accommodation that makes the Aztec ideology of space, time, and human history so powerful and effective.

Back in the turbulent 1960s, an era of youth-centered rebellion against institutions, some of my students hatched a diabolical plan to deactivate and remove from campus all the centrally controlled school bells that noisily rang out time signals to begin and end classes. I can even remember the low, buzzing sound that used to precede each signal by two minutes to warn instructors to begin to drop their voices and to suggest to students—as if any hints were needed—that the time to close their notebooks was approaching. Stuffing the bells, which were located quite accessibly in the middle of each hallway and barely six feet off the floor, with rags had already failed; moreover, doing so was deemed a fire hazard. So, in a single night and with only a few screwdrivers and pairs of scissors to snip the connecting wires, the naughty students put their caper to destabilize time's rule into full operation. It came off a complete success. The culprits were never caught nor were the bells ever replaced. I have often wondered whether this institutional inaction resulted from a detached administration or one that realized that mid-1960s education perhaps had become too regimented. Since then no bell—save the one in the nondenominational chapel at the center of the campus, which peals only at 10 o'clock on Sunday morning—has ever tolled again at the college. Paradoxically, classes seem to run as close to on time now as they did in the pre-gong era!

Indeed time does rule life.[18] We have seen it in the way the ancient Aztecs captured time's controlling power in the axis of their ancient capital of Tenochtítlan and in the way mercantile power brokers regulated life in the medieval city with the "Big Bens" that marked their centers—a feeble remnant of which I still hear on quiet nights emanating from down in my village. But in both cases religious ideology lay at the foundation of timekeeping in the city. Our classroom bells are a lasting memento of old Saint Benedict's contribution to marking time.

Although the signals tolled out by bells and the delicate interplay of silent light and shadow in the city are now largely meaningless—perhaps only appreciated by romantics—time's grip on urban life is ever present. Cultural globalization has already made it a universal phenomenon. Time's beat has become individualized in the little Big Bens we wrap around our wrists. Their private time beeps out within earshot of the wearer (and all too unfortunately to the proximate public). Time in and time out have been relegated to the sports field but are rarely punched on the worker's clock. Labor's bells have been silenced by flex time, and quantified work output now seems more the metric of duration. What we do is often overseen by a supervisor who is more distant from our computer terminal than one medieval city was from another. Those of us who migrate to and from a central urban workplace face two rush hours far longer than any of the other hours that make up the day, including the abbreviated lunch hour and all but extinct happy hour.

Rather than answering a centrally located work bell to start our day, we all rise up to the sound of the personal bedside alarm, which initiates our carefully timed, swift journey through bath and (maybe) breakfast, then on to the short drive to the train station (when was the last time you looked at the clock in the terminal?) or the longer commute down the urban interstate ribbon. Modern city life warps our old ideas about time. Multitasking has replaced the sequenced work activity of the Benedictine Order.

Einstein was right—although he never intended to apply his relativity principle to commuting—time and space *are* inextricably interwoven. Here and there in the city are only as far apart as time will allow—either a $10 or $20 cab ride depending on the time of day you step into the vehicle. Time marks long-term life in the city as well. Just as many of Tenochtítlan's seventy-eight stone temples once commemorated Aztec mythic time (I told you only about the main one), memorials in our cities, known as museums, mark events that dissolve into myths to be told in the future. They celebrate the accomplishments of the times of each president, the landing of the pilgrims, and the advent of rock and roll. America also boasts museums dedicated to somewhat less mythologically or historically important figures and ideas like Liberace, Lizzie Borden, Pez Candy, Barbed Wire, UFOs, Prostitution in the Old West (the Julia C. Balett Red Light Museum in Virginia City, Nevada), Bad Art, and Questionable Medical Devices. In a town not too far from where I live there's even a cheese museum, which pays homage to a time when upstate New York once richly sustained itself as a dairy-producing area. Some time temples are dedicated to events that scar our memories so deeply that they must never be forgotten—the recently designed World War II Memorial, the Vietnam Memorial, and the Holocaust Museum. More recently we've developed a penchant (some call it an obsession) to seize not intervals but singular moments in cataclysmic time. The Memorial at the Murrah Federal Building in Oklahoma City and the much anticipated memorial to the tragedy at the World Trade Center spring to mind.

Of the many designs for the new World Trade Center one of particular relevance to the subject of this chapter—how we manage the meting out of time—stands out. Architect Daniel Liebeskind has come up with a plan for a building complex scheduled to be completed in 2013 at a cost of ten billion dollars. Its main tower will be, aptly enough, 1,776 feet tall. I am impressed with the way it captures the moment. King Moctezuma would have loved it!

> To commemorate those lost lives I have created two large public spaces, the Park of Heroes and the Wedge of Light. Each year on September 11 between the hours of 8:46 AM, when the first airplane hit, and 10:28 AM, when the second tower collapsed, the sun will shine without shadow, in perpetual tribute to altruism and courage.[19]

Oddly enough, in May 2003 an architect who studied the plan claimed that the Wedge of Light would actually be in shadow because of an intruding building across the street. At this writing Liebeskind's dilemma is unresolved. But even if the builders make mistakes, the World Trade Center Memorial is still based on architectural principles designed to maintain our connections with the cosmic beat and to each other.

NOTES

1. Peter Wilson, *Domestication of the Human Species* (New Haven: Yale University Press, 1988), 75.

2. Numa Fustel de Coulanges, *The Ancient City* (New York: Doubleday, 1873), 160.

3. Jacques Le Goff, *Time, Work and Culture in the Middle Ages* (Chicago: University of Chicago Press, 1980), 43–52.

4. Benjamin Franklin, "Advice to a Young Tradesman, Given by an Old One" (Philadelphia: New Printing Office, 1748), 1.

5. Codex Mendoza, Bodleian Library, Oxford, manuscript, Selden Archives (HMAI Census No. 196), A, i:62v.

6. Juan de Torquemada, *Monarquia Indiana* (Mexico City: Porrua, 1961 [1723]), vol. I, book 2, chap. 44, p. 188.

7. Ibid.

8. Miguel León Portilla, *Aztec Thought and Culture: A Study of the Ancient Nahuatl Mind* (Norman: University of Oklahoma Press, 1963), 166.

9. Bernal Diaz del Castillo, *The Discovery and Conquest of Mexico, 1517–1521,* ed. Irving Leonard (New York: Farrar, Straus and Cudahy, 1956), 201–202.

10. Hernan Cortes, *Letters from Mexico,* tr. and ed. Anthony Pagden (New Haven: Yale University Press, 1986 [1519–1526]), 105 (second letter).

11. León Portilla, *Aztec Thought and Culture,* 436.

12. Bernardino de Sahagun, *Florentine Codex: General History of Things of New Spain* 3, tr. A.J.O. Anderson and C. E. Dibble, *Monographs of SAR* (Santa Fe: School of American Research, and Ogden: University of Utah Press, 1978 [1585]), 1.

13. Ibid.

14. William Wordsworth, "On Seeing the Foundation Preparing for the Erection of Rydal Chapel, 1823," in *Wordsworth: Poetical Works,* ed. Ernest de Selincourt (Oxford: Oxford University Press, 1969).

15. Toribio Motolinia, *Memoriales, o Libro de las Cosas de la Nueva España y de los Naturales de Ella,* ed. Edmundo O'Gorman (Mexico City: Instituto de Investigaciónes Historicas, UNAM, 1971), 51.

16. I am fairly convinced that this is how they did it. There is little reliable direct evidence that the Aztecs cared about a leap year.

17. We know that the calendar by count would fall short of the alignment by about 0.25 days per year, or approximately 5 days in 20 years, 12.6 days in 52 years; therefore, as measured by the 365-day count, the equinox sunrise would shift 5.5 arc minutes to the south (right, or clockwise if you view it from above) per year, or 55 minutes in 10 years, 5°16' in 52 years. Since the diameter of the sun's disk is 30 arc minutes, it becomes fairly obvious that it would have taken scarcely a decade to recognize that the alignment was not holding up.

18. The motto of the National Association of Watch and Clock Collectors.

19. *New York Times* on-line www.nytimes.com (May 1, 2003).

SEVEN

SPACE: ARE MAPS REALLY TERRITORY?

[I]n that Empire, the craft of Cartography attained such Perfection that the Map of a Single province covered the space of an entire City, and the Map of the Empire itself an entire Province. In the course of Time, these Extensive maps were found somehow wanting, and so the College of Cartographers evolved a Map of the Empire that was of the same Scale as the Empire and that coincided with it point for point. Less attentive to the study of Cartography, succeeding Generations came to judge a map of such Magnitude cumbersome, and, not without Irreverence, they abandoned it to the Rigors of Sun and Rain. In the western Deserts, tattered Fragments of the Map are still to be found, sheltering an occasional Beast or beggar; in the whole Nation, no other relic is left of the Discipline of Geography.

—JORGE LUIS BORGES[1]

Lewis Carroll, creator of *Alice in Wonderland,* tells the story of an imaginary cartographer who experiments with making maps of his country at a larger and larger scale until finally he arrives at a one-to-one representation of his country. But the map is impractical because it can't be unfolded. If it were, it would cover all the arable land and even block out the sunlight. So, Carroll tells us, people learned to use the country as a map of itself. Perhaps Lewis Carroll was the source of writer Jose Luis Borges's commentary in my epigraph, which was intended to poke fun at the habit Western culture has acquired of making exact large-scale models of things: space stations, theme parks, world's fairs, the failed Biosphere, and the endless virtual environments conjured up in the cyberworld.

In the eyes of some, the virtual image, even though it means "being in essence if not in fact," can seem better than equal to the real article. I vividly remember a trip to the site of a rare total eclipse of the sun in the Caribbean a few years ago. I was on a cruise ship, assigned the pleasant task of lecturing a group of amateur astronomers on eclipse phenomena. While most of us followed the partial phases of the eclipse in blazing tropical sunlight out on the open deck through the dimming glass of welder's helmets, one member of the group sat in his air-conditioned stateroom, monitoring on his computer the slow progress of the moon across the disk of the sun. As the last visible crescent thinned, casting a pale yellowish hue over the sharp-shadowed landscape, we pled with him to come out on deck and see the phenomenon reach its climax. Then the diamond ring appeared—the last flash of sunlight from the jagged edge of the black lunar disk—followed by a sudden plunge into total darkness. "Come now! Totality! It's happening!"

For three minutes and fifty-eight seconds we witnessed one of nature's most wondrous celestial spectacles—darkness at noon. We were among the fortunate to be situated in the pencil-thin shadow that scratches out a line of momentary daytime darkness across the globe. The sun turned black and revealed its attendants: Venus and Mercury; the stars came out and the wind picked up as the temperature plummeted several degrees. Seabirds vocalized wildly and some passengers shed tears of joy at glimpsing this rare, hidden side of nature. All the while our room-bound, cyber-sun-gazing cohort followed the event in a universe entirely unto himself on the laptop screen; he said he couldn't get away from it; he said he could "see it better" this way (how would he know that?). But he didn't *see* it. He saw a representation of the eclipse—an amplified video image electronically beamed through an eight-by-ten-inch translucent window. He escaped the context of the event: the panorama of the whole sky, the sound of the wind and the birds, the human voices responding to the extravaganza. He missed the experience. He traded in the thing for an abstraction of it. He saw a *map* of an eclipse.

Traditionally, we think of maps as graphic representations that help us understand things spatially. Maps are reduced substitutes that we invent to give us an idea of the real thing. We draw them to scale, usually on a flat surface. The word *map* comes from the Latin word for tablecloth, because the earliest maps of the Western world—the Romans called them *mappaemundi*—were large, foldout cloth documents. When they represent places, maps are almost always oriented with north at the top and east off to the right. In earlier maps east used to be at the top because it was the place worshippers faced when they prayed to the rising sun. (Recall our discussion of the word *orientation* in the previous chapter.) East was also the direction to Jerusalem in the Mediterranean-centered world.

I remember when I drew my first map. I was five and I guess I felt a need to represent the world I had experienced. My map was a sky view of the world surrounding my house at 156 Morris Street, next door to the Truman Street School in the Hill District of New Haven, Connecticut. I positioned my house in the center of the map, at mid-block on the right (west) side of a square bordered by other houses about the schoolyard. Four streets—Morris, Truman, Barclay, and Washington (technically an avenue)—bordered my square. I recall drawing other squares on each of the four sides of my own, separated by these streets. Then I labeled the houses on the squares with the names of their residents. I knew the names of those who lived in close proximity to the center of my world map, but as I ventured farther from the center the gaps in my knowledge grew.

Pedaling over vast distances by tricycle wasn't an extensive means of acquiring cartographic information for a curious five year old (especially one with a very protective mother), but I did manage to place four more squares kitty-corner on my map, for a total of nine in a three-by-three grid. I even located a tenth square at the mid-bottom of the sheet (at the astronomical distance of 2½ blocks from home base!) because that's where my paternal grandmother lived and I frequently walked there to visit her, through a no-man's-land of two- and three-story houses. Those ten squares pretty much constituted my experiential world until I was able to expand my sphere of inquiry by acquiring a two-wheeler. When I was old enough to cycle several blocks to the library, I learned that just about every other urban kid's universe looked almost exactly like mine: compact squares in a grid, some with beveled edges, others split diagonally—necessities forced on urban planning by natural contours in the landscape, such as hills, streams, and, in the case of New Haven, an estuary-laden coastline.

Even the center of my home town mirrored this pattern. English Puritans planned it that way when they founded New Haven Colony in the 1630s. When I zoomed in on the map of downtown in the library's atlas, I discovered a version of my little packet of squares almost exactly the way I drew it. There were squares abutting other squares; nothing but squares all across the landscape. Flipping through the atlas, I found page after page of even bigger quadrilaterals connecting city and town, hill and vale, state by state through the countryside, across the wheat and cornfields of the Midwest all the way to the ocean shore on the other side of the continent. These squares, I learned, were part of a colossal grid that wrapped the world.

Why do we represent the real world in such an unreal way—a ball with an axle through it and a chicken-wire grid laid on top of it? To answer that question, you need to do more than browse an atlas. The explanation has as much to do with who we are as a culture as it does with the environment we're trying to conceptualize. I want to suggest two things here. First, everyone maps the real world and there are vast numbers of ways of mapping it. Our way is but one of them; it is peculiar to us and, like our version of the solar system described in Chapter 2, it is neither more nor less a "true" representation of reality than the rest. And second, our way of picturing the real world has evolved out of conscious actions and choices we've made in the past. The story behind these decisions has many sinuous turns. Had we elected to follow a different path at any point, our modern maps in the Western world could have turned out quite different.

To follow some of the major bends and turns let me transport you over some of the cartographic turf walked by our ancestors—the people of the Aegean, the world of Homer and Aristotle, the Romans and their European descendants who lived in medieval times and the Renaissance. To discern just how peculiar our view of the world is, we'll also need to reflect on how other cultures, some now long extinct, chose to represent the world they experienced.

Few of life's really big questions have clear-cut answers. So we shouldn't be surprised to learn that experts the world over still dispute the question, who made the first map? Arguably, it was a resident of Çatal Hüyük in Turkey around 6200 BC, who painted a representation of his town on a wall. The map pictures houses of a settlement at an oblique angle, as if viewed from a hillside that overlooks the town. Rock carvings from the Bronze Age (2500 BC) in Italy look a lot like plans that depict places of residence. The same is true of rock art in Mexico and Australia, which may date to even earlier times. But these artifacts may be pictures rather than maps. They use pictorial icons. They don't look like plans; they don't have a bird's-eye point of view of a landscape—what *we* would call a map. Regardless, they do suggest that very early people were using different media to express their location relative to the world around them.

The ancient city of Nippur (Figure 19), which flourished in the middle of the second millennium BC in the fertile land between the Tigris and Euphrates (about midway between Baghdad and Basra in today's Iraq), has yielded a document that no one could possibly doubt is a map. It rains heavily in the spring in Iraq and when the rains end, the unrelenting summer sun dries out the landscape. Before irrigation the untamed land suffered massive annual flooding followed by extreme desiccation. The flood plain that bordered the Tigris and Euphrates became caked with mud-clay. Because they are indelible, large, flat hunks of dried clay are ideal writing surfaces. Once scratched with a metal tool or hammered with a stylus bearing a wedge-shaped point, a message in clay can be preserved virtually forever. That's how cuneiform (named after the *cunei,* or wedge, shapes on the tool), the earliest writing system, was born. A clay tablet map from Nippur bears cuneiform writing. It is part of a labeling scheme attached to a map of the walled city. You can see a temple plan at the bottom, a canal in the middle, and storehouses at the top—all so labeled. Outside the multi-gapped wall are roads, canals, and a moat. The Nippur tablet describes that particular landscape, and only that landscape. A slightly earlier plan dated 2300 BC from the city of Nuzi is a bit less graphic, but some cartographic historians have given it precedence over the Nippur tablet.

Most of these early maps in clay are related to the delineation of territory. Broken cuneiform tablets found in ancient cities' trash heaps mined by archaeologists deal with property transactions, land disputes, and estimates of grain output; a walled domicile with storehouses and an irrigation canal running through it turns out to be the subject of the Nippur map. In fact, the merchants and tradesmen of the Fertile Crescent, as historians call the Middle

19. This map of the walled city of Nippur, dated to about 2000 BC, is one of the earliest known maps. (Friedrich-Schiller-Universität Jena)

East, produced so many bills of lading and account books that discards were used as rubble fill in the construction business.

Some of these early maps in clay suggest that Mesopotamian thought was not completely focused on business and commerce.The first Mesopotamian world map, dating to about 600 BC, shows Babylon at the center. The mythic lands of heroic tales and adventures on the salt sea (the Persian Gulf) are represented by triangles jutting out of the circle outlining the rim of the visible world. Greek geographer Strabo (who lived during the first centuries BC and AD) credits ninth-century BC poet Homer with founding the science of geography, for

> Homer has surpassed all other men in his acquaintance with all that pertains to public life. . . . And this acquaintance made him busy himself . . . about

the geography both of the individual countries and of the inhabited world at large, both land and sea.[2]

I am struck by the focus on life and habitation in this rather flattering passage. Like all good geographers of the Western world, Strabo and Homer were concerned with the problem of the structure of the *oikoumene,* the management of the household or the state (whence our word *economics*). They wanted to represent the habitable world—the territory of *human* occupation.

The world revealed in Homer's *Odyssey* may seem very different from ours; but 3,000 years later, as we plumb the depths of the universe in search of intelligent life, we continue to ask many of the same questions Homer asked. How far does the world extend? Who lives there? What determines its limits? Like the Mesopotamian map and my own, the Homeric world map had home base at the center and all travel routes were arranged according to a home-centered coordinate system. On the Greek map the world was represented by a flat disk—circular (remember Anaximander's hockey puck in Chapter 2?), oval, oblong, or even rectangular—made up mostly of a single landmass surrounded by water. The outer water was a constantly moving ocean-river they called Oceanus. The high vault of heaven, often represented by an inverted hemisphere, rested on the rim of the disk. The sky was supported by a series of tall, invisible pillars cared for by Atlas or Hercules. Gibraltar, at the western terminus of the Mediterranean, was once called the Pillars of Hercules. Some even said that Hercules's broad shoulders actually supported the sky.

According to Homer and later geographers Oceanus ebbed and flowed, a gentle swell going nowhere really. In fact, the tide in every inhabited region proved that there was one, and only one, ocean. The early Greeks made no distinction between the outer and inner sea, the ocean and the Mediterranean; it was all one and the same to Homer. Strabo agreed that the limits of the inhabited world were washed by the sea because, he reasoned, our senses tell us so; for no matter in which direction man has traveled, the sea has been found, just as our senses today support the notion of a vast, unlimited extent of space within and without, above and below us. Water not only surrounds but also underlies and overlies the *oikoumene.* It falls down on us from above as rain and seeps up from the ground below in artesian springs.

Hyperion, the great sun god, rose daily out of Oceanus on one side of the earth disk and every night sank beneath the waves on the other side, "drawing black night over earth, the grain-giver" (in Homer's words).[3] Some who had ventured far out to sea even swore that they could hear the hissing roar of the fiery ball of the sun as he plunged beneath the water at sunset. But no one knew exactly where Hyperion came from or where he went to dry out at night.

Regardless of what you believed about the shape of the earth, you couldn't talk about a map of land and sea in ancient times without telescoping your model upward and outward to encompass the sky. The stars followed the course of the sun, traversing the heavens "after having bathed in Oceanus."[4] All except the Bear. "She alone hath no part of the baths of Oceanus," says Homer, as she wandered around the fixed pivot of the heavens marked by the eternal Pole Star.[5] We have DVDs, the Internet, and iPods, but the sky was Homer's medium for telling stories. He told star stories as the constellations slid by overhead, stories of life featuring "the late-setting Bootes" (a herdsman), who took to the plow to get ready for the planting season.[6] And in the hunting season there was mighty Orion (a hunter with a club). The sky map imitated the map of the earth. It, too, was segmented into squares.

Those ancient world maps also marked the winds. Homer lists four of them. Boreas, the north wind, was a stormy petrel bringing sudden squalls and very dangerous to navigators. Zephyrus, the west wind, was often represented as a stormy wind, but not by Homer who knew the west, where the climate was temperate and the people were prosperous. Also in the west were the Elysian Fields and the ends of the earth, where life was easy. As Homer says, "No snow is there, nor yet great storm; but always Oceanus sendeth forth the breezes of the clear-blowing Zephyrus."[7]

The logical order that our ancestors created in their water-bound *mappaemundi* satisfied them and it would continue to work for a long time. An explicit attachment between heaven, earth, and water exists in this ancient concept of the world, a cosmic wholeness that made common sense to them because it aligned with their experience.

These old maps look warped to us; sizes and distances appear distorted when compared with those of Rand McNally. On Hecataeus's sixth-century BC map of the *oikoumene* (Figure 20) the Iberian Peninsula is too small, the Arabian Gulf too long. Italy is stunted, Sicily too large. (No surprise here; Sicily was home base for most of the early mapmakers.) Distances were mere estimates based on the time it took to travel by land (where mountains made

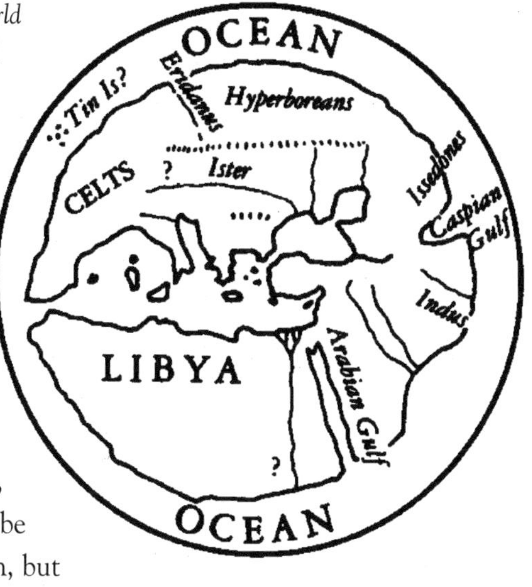

20. Hecataeus's sixth-century BC world map shows recognizable, if distorted, land features. (From Oswald A.W. Dilke, Greek and Roman Maps *[London: Thames and Hudson, 1985], 56)*

gauging distance difficult) and by sea (where the wind was a factor). To make matters worse, there was no universal agreement on the distance unit, the *stadium.* It was supposed to be the length of an Olympic stadium, but which one?

These early maps just don't depict the kind of order in a representation of the world that makes sense *to us.* We like our order to be expressible in terms of latitude and longitude. We want it to be scientific and precise. Times Square in New York, New York, is 40°25' north latitude, 73°59' west longitude (and 212 miles by road from my front door—even if traffic jams temporally lengthen the last 20 miles). But now that we understand their common sense and are aware that they were our ancestors, we need to ask, how did our new and different kind of common sense that gives us our map of the world arise? What changed?

Our world is like an orange. We slice it into horizontal bands (latitude lines) separated by the tropics and the equator, and vertical ones to segment the globe longitudinally. This zone model of the earth had its origin in the work of the great astronomer Ptolemy of Alexandria, who lived about the second century AD (we met him briefly in Chapter 2). Ptolemy would give us in his great work, *The Almagest,* a vast compilation of star positions—more than a thousand of them—along with a theory explaining their motion about a fixed earth that would make sense for 1,500 years, all the way to the scientific Renaissance, when Copernicus's heliocentric (sun-centered) theory would replace it.

Ptolemy's concept of hierarchical order stems from Aristotle's fifth-century BC idea of a geocentric (earth-centered) universe with all matter arranged around it in neat layers, one on top of another. The sphere of the heavens lay on the outside, the sphere of the earth on the inside, with spheres of water, air, and fire sandwiched between. Stones fall to earth; air bubbles up through water; rainwater falls toward earth from above; fire licks its way through the air. It all makes perfect sense: everything seeks its natural place in the hierarchy of matter; the heaviest stuff goes to the bottom, and the lightest rises to the top.

In the zonal model, the space of grass and trees, desert and tundra, is made to conform to a series of stratified bands that run across the basically east-west oriented, rectangular Mediterranean basin. In the framework of this world map, the earth is segmented into latitudinal zones or climates (*climata*). The latitude, or distance from the equator, was calculated from the length of the longest and shortest days based on the angle of the noonday sun (*clima* actually means "inclination"). As the mapmakers' informants ventured ever farther from home, it became clear that the weather of a place—its temperature and precipitation—had something to do with its distance from the equator, or its *climate.* It also became clear that the globe was round in shape, which was obvious when observing the varying angle of the noonday sun on a given day from the overhead position (the zenith) as you traveled to different latitudes. So Ptolemy used measurements of the sun to divide up the earth. Conceptually he was the first person to leap off the surface of the earth to a new vantage point from which to conceive a graphic representation of the world. We can appreciate why our predecessors couldn't really detach heaven from earth in their thinking. Heaven and earth, life and the water that nurtured it, were all part of what we could call a world view or a vision of the entire cosmos, much, much more than just a practical map showing how to get from here to there.

You pay a price when you stick your neck out above the world the way Ptolemy did. If you want to represent the shape of the world as you think you really know it to be—a sphere—then you need to move your eye over a huge globe to find where home is. And it must be a very big globe to contain the names and locations of all the places that matter—the ports we sail to, the cities we trade with, the exotic beyonds that tease our speculative imaginations. On the other hand, should we choose to represent the real world as flatland, at least we can anchor our viewpoint locally, even if we do so at the expense of denying the earth its "real" spherical shape. We need to make a choice.

For Ptolemy, reconciling the two-dimensional world of rectangles with the three-dimensional globe and its band of *climata* exacted another price: distortion. Try folding a parachute or a set of custom-fitted sheets and you'll see the problem. Now imagine using your contorted bed sheet as a device for representing distances between real places. Check out a Mercator projection map in any atlas and you'll bear witness to the outcome of the long-lasting Ptolemaic compromise—northern Canada and south Australia are way too big, northern Greenland and all of Antarctica absurdly huge. When you replace one sort of representation of reality with another you pay a price.

Finished about AD 150, Ptolemy's *Geography* (like *The Almagest*, another groundbreaker) was a massive work, the first to put mapping on a truly scientific plane. It is imbued with Aristotle's notion that in this seemingly chaotic world, everything has a logical, natural place and that in a stable, static universe (the only kind of universe one could imagine at that time) everything should be in its proper zone or *clima*.

Ptolemy's work has had a profound influence not only on how we represent our physical world 2,000 years later, but also on our early thoughts about others who share the globe with us. When it came to the remote regions, Ptolemy based his model of the world more on reported itineraries of travelers than on measurements of the sun. Take the distant land of Agysimba, a far away place where black people dwelled. According to Aristotle's unquestioned concept of order, animals and plants most resemble animals and plants in other regions with the same temperature and atmospheric conditions—in other words, at the same parallel of latitude. Now Ptolemy knew that black people, rhinoceroses, and elephants were not found in his hemisphere until one passed into Ethiopia, below the northern tropic; he also observed that as one moved closer to the equator, the human complexion darkened. Following this logic, Ptolemy placed Agysimba at the same distance to the south of the equator as Ethiopia lay to the north. We could say that he had a hemispheric bias. He used things in his own hemisphere to define what ought to be true in the other. He put people in their "proper" place to conform to his physical model of the world. Such is the power of the rule of symmetry.

Aware that the circumference of the globe was much larger than all the known world of Europe, Africa, and Asia that had been explored, Ptolemy realized that there was plenty of space on it for other inhabitants. Thus, he made the assumption that they should differ little from the race of the *oikumene.* (His assumption is akin to the one most contemporary seekers of extraterrestrial life seem to adopt.) Ptolemy's love of the sphere's symmetry likely led him to the notion that we of the *oikumene* (in the old sense of that word) inhabit one of four symmetrical zones—two in the northern and two in the southern hemisphere—separated from each other by four oceans. Quarter the earth-orange and you get the idea: the quarters of the world become islands forever out of communication with one another. All of this is reminiscent of our earlier discussion of the number four.

The Romans vastly expanded the early Greek oblong representation of the world. The first-century AD emperor Augustus had an interest in founding colonies and providing land for discharged Roman civil war vets. He would kindle a vast empire that ruled the world from Iberia to India. But getting east-west locations pinpointed in a basically east-west stretched civilized world posed a practical problem. North-south is a snap because, as you'll recall, you can determine latitude either by sighting the altitude (in degrees above the horizon) of the North Star or that of the noontime sun. But nailing down an east-west position requires a time standard, because that's the way the world turns—along the direction of the sun's path on the sky. Put simply, the difference in longitude between two places is the same as the difference in time. Folks on the central meridian of the Pacific Standard Time Zone (the 120th) see the noontime sun exactly three hours (45°) later than those who live on the 75th meridian in the Eastern Standard Time Zone. (Since there are 360° in the earth's circumference and 24 hours in a day, the time it takes the sun to traverse the globe, one hour of time equals 15° longitude.) Precise time yields precise longitude. But the technology that could produce ship's clocks, those that could function in the rolling swells that accompanied voyages across large bodies of water (forget pendulum clocks!) didn't develop until surprisingly late—the early eighteenth century, in fact.

The Romans transformed the Greek rectangular grid to a square grid and they carted it all over the empire. Before anyone built anything, much less moved in, the *agrimensores* ("land surveyors") showed up with their *gromae,* cruciform instruments that look like upside-down ceiling fans on long poles, with plumb bobs hanging from each of their four tips (Figure 21). The surveyors blocked out towns into *insulae* ("islands") and the fields around them into

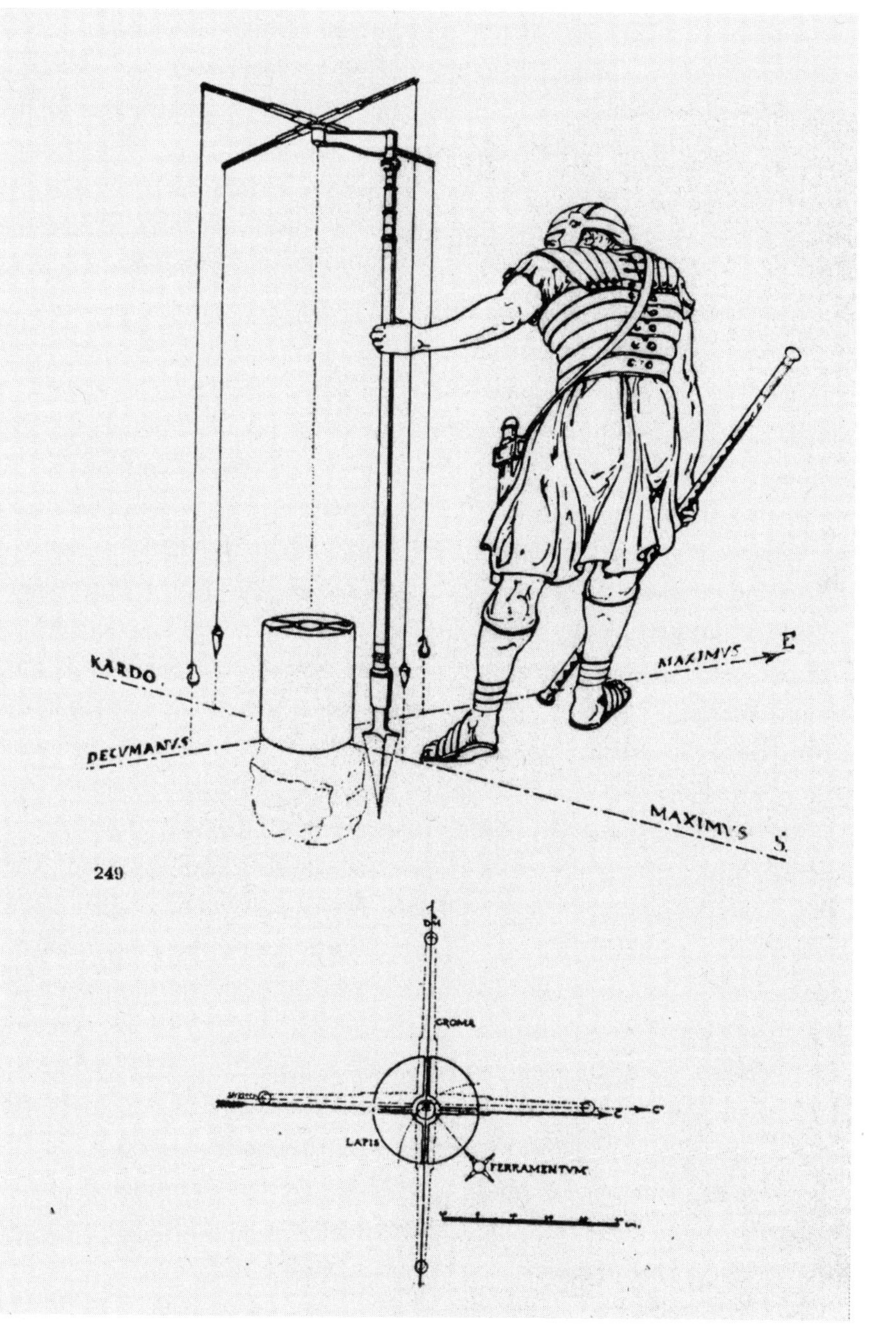

21. *A Roman* agrimensor *lays out the major axes of a town. (Drawing by P. Frigerio,* Antichi Istrumenti Technici *[Como, 1933])*

centuriae ("centuries" consisting of 100 squares) of fixed dimensions. Now I know where I got the idea to make the map of my New Haven neighborhood; I had followed the Roman plan of centuriation. A good eye can still pick up the mark of the *agrimensor* on European maps and aerial photographs. Recall (from the discussion in Chapter 5) that the two main roads that intersected at the center of town were called the *Decumanus* (the north-south street) and the *Cardo* (east-west); *Cardo* means "hinge." It is as if you can fold up a replica of your town along the east-west axis like a Ouija board and carry it off under your arm. Blocks are labeled (in Roman numerals of course) to the left (west) or right (east) of the *Decumanus*, and to the near side or far side of the *Cardo*. The viewpoint seems to be situated somewhere up in the sky south of the overhead position.

If Greco-Roman maps deal basically with finding your way through territory and inventorying the world both within and without the familiar *oikoumene*, medieval maps reflect changing values. They take on a decidedly religious, more humanistic quality. The sky sphere gets separated from the earth sphere, which Christian theology seems content to have abandoned, attaining greater satisfaction by giving a central position to the Holy Land, for as it is written in the Bible: "This is Jerusalem; I have set her in the center of nations with countries round about her."[8]

The medieval European world is represented on maps of various shapes: irregular, ovoid, rectangular, clock-shaped, and wheel-shaped. One plan shows a box-shaped world, its proportions like the Ark of the Covenant (dimensions: 2½ x 1½ x 1½ cubits). Here is a world distorted metaphorically from the rational, precise, one-to-one reality that appealed to the Greeks—and still appeals to us today. The Holy Land is magnified in size, and the word of Scripture trumps observation and reason. In many of these maps the world's land masses are distributed in the shape of a T, cut one way (usually) by the Mediterranean and the other way by prominent rivers. These T-O maps (Figure 22) as they are called (the "O" part comes from the surrounding Oceanus) partition the world into Europe, Asia, and Africa, the remnant lands that surfaced after the flood, each said to have been repopulated by one of Noah's three sons.

Medieval maps stress a different kind of unity—not the unity of the domains of water and land or land and sky, but rather that of man and God. Their makers were trying to represent the everyday world in the context of a differ-

22. The world is what you make of it. This medieval T-O map shows a world divided into the shape of a cross. (J. B. Harley and David Woodward, eds., The History of Cartography *1 [Chicago: University of Chicago Press, 1987], 297, fig. 18.4)*

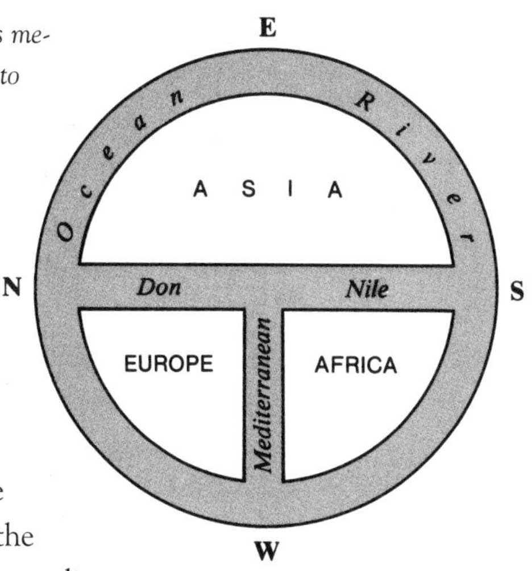

ent kind of truth—a truth acquired by revelation rather than by physical exploration. The maps reflect the power of religious belief, especially the high value placed in the covenanting religion of Christianity. We find Adam and Eve—even the snake who tempted them—on some maps. On others, paradise becomes the distant continent to the east, the land visited only by Christ's apostles. But surviving and still prominently displayed are the circumfluent ocean and the four winds, 2000-year-old remnants of Homer.

Some of the more exquisite medieval maps were up to five feet long and drawn on vellum or cowhide. Unlike modern maps they were not carried in the accessory compartment of a horse-drawn coach. Rather, they served as altarpieces or were hung behind the altar of a church. For example, the Hereford world map, ca. AD 1100, is based on classical itineraries and later sources, a mix of the secular and sacred geographical lore of the Middle Ages. Mythical places are depicted, but the Hereford also contains (especially in the European section) new information derived from medieval commercial journeys, pilgrimages, or Crusades. In its coloring and adornment, as well as in its view of what the real world was conceived to be, the Hereford map of the world and similar maps beautifully express the feeling of the immanence of the coming of Christ's kingdom on earth. The map, like the world, is more than mere territory.

By the time the scientific Renaissance began to flower in Europe in the sixteenth century, few favored anything but a representation of a spherical world. And most found great advantage in displaying the sphere as a secular rather than a religious one. A literate tradition of geographical knowledge began to develop, accompanied by a new technology that included the compass and the printing press. Columbus read Marco Polo and Captain Cook would read Columbus. Still, the pre-1492 globe showed nothing but a vast expanse of ocean between Europe and Japan—an impression that would soon change as the Americas came into focus. That old sense of balance and symmetry inherited from the Greeks had led explorers to search out the imagined distant regions of the earth—the unknown land directly opposite the *oikoumene* at the other side of the world (the "antipodes"). It was conceived as a sizeable continent, assumed to occupy a large portion of the southern hemisphere—and the search for it was one of the principal motives behind Captain Cook's voyages.

As had been the case with Ptolemy, the more remote the frontier, the odder one imagined the inhabitants to be—peculiar beasts and strange races (e.g., men with eyes in their chests and a tropical tribe with extra big feet they could extend over their heads, the better to shade themselves from the blazing tropical sun) (Figure 23). (Do we imagine any less exotic extraterrestrials in the present age?) Such fantasizing gave birth to the idea of the lost continent of Atlantis. By the middle of the sixteenth century the assumption of an inhabited antipode, coupled with the appetite for precious goods, produced an expansion of Western knowledge of the world unparalleled in human history.

The great voyages of discovery coincided with the conceptualization of a New World perspective that placed the earth in motion about the sun and the sun among a limitless sea of shining stars. For the first time one could conceive of an incredible vastness not only of the earth we inhabit but also of the universe. Our desire to conceive of things in spatial terms (derived from the Greeks) along with our tendency (like nature) to abhor a vacuum, has led us to explore, map, and populate all existing spaces with our own kind, from our home country to the moon, Mars, and beyond. We haven't stopped yet, for who doesn't believe that spaceship earth is ours to pilot across the vast waiting intergalactic ocean that surrounds it?

Guided by changing values, our Western way of mapping the world has changed drastically since antiquity. My sweeping tour of our kind of map's origins to represent it has scarcely done justice to a deep understanding of all that has happened in the exciting story of mapmaking. Nonetheless, if you look at any map today, you'll recognize the influence of the impassioned Greek geometer, the practical Roman agrimensor, the outward-reaching Renaissance

23. *The more remote the place, the stranger its inhabitants.* (Strange Races, *Model Book,* MS *Harley 2799, f. 243, German, twelfth century, by permission of the British Library)*

cartographer, and the indefatigable European explorer and colonizer. Our maps could have been made by no one but ourselves. Change one aspect of one of the source cultures that contributed to the enterprise and our maps become vastly different. How different? You can't even begin to answer that question without looking in depth at somebody else's maps—maps created by people in total isolation from Western-based culture, maps fashioned by people no less intelligent than our forebears who lived on the other three quarters of Ptolemy's globe. My choice of the three kinds of maps to tell you about is biased by my own special interest in the people who made them.

Water flows like traffic merging on the interstate. Paddle your way up any river and you'll confront a series of V-shaped headlands at points where tributaries join the mainstream. Travel north on the Mississippi just across the Louisiana-Mississippi border and you'll see the Homochitto veer right. One hundred miles farther up, the Big Black makes its angled entry; north of Memphis it's the Loosahatchie, then the Hatchie, the Forked Deer, and the Obion. Sail up any one of these tributaries and you'll find an even finer reticulated pattern of conjoining waters. In particularly wide valleys (e.g., the Orinoco, Nile, or Mississippi delta) the main flow path splits into a branch pattern, the spaces between carved into wedge-shaped lands looking like so many arrows that point to their departure point in the middle of the valley.

That's what the landscape is like where the Huatanay and Apurimac rivers join in the valley of Cusco in highland Peru. Descendants of the Inca call these junctions *tinkuy* in their native Quechua language. *Tinkuy* means "the harmonies that balance, or the blending of opposites"—where things come together. There, at the junction of the two major rivers, the Inca built their capital city over 500 years ago. Today Cusco is the jumping off point for adventurous tourists determined to hike the Inca trail to Machu Picchu. Five hundred years ago it was the stronghold of an empire that has been compared to that of Caesar or Alexander, a domain carved out of rugged river-gorge terrain in the high Andes that stretches from Ecuador to Argentina.

As you might have surmised when we met them back in Chapter 5, the Inca were a very well organized, highly creative people—so much so that they devised a map of their world that was as imaginative as anything Borges or Carroll could have conceived. It was the very city they lived in. The Inca *ceque* system, so reminiscent of the Borges quotation I cited in my epigraph, was a gigantic, invisible mnemonic device built into both the natural and man-made topography of the Inca capital. Of it, the sixteenth-century Spanish chronicler Bernabe Cobo wrote:

> From the Temple of the Sun as from the center there went out certain lines which the Indians call ceques. . . . On each of these ceques were arranged in order the guacas—shrines which there were in Cusco and its district, like stations of holy places, the veneration of which was common to all.[9]

There were forty-one of these *ceques*; you can think of them in the same way as the imaginary lines of latitude and longitude on our maps, except that they all go out from a point (more like what mathematicians would call a polar, as opposed to a Cartesian, coordinate system). And (this is important) people actually used them in everyday life. Water was important to the Incas. In a montane environment this precious commodity can be wasted or lost if it isn't properly channeled. The *ceque* lines marked out the water rights of the various kinship groups that formed the capital's population. They believed themselves to have descended from ancient ancestors who lived inside Pachamama, or mother earth. It may seem odd to us to locate places in a radial mapping scheme, but it makes perfect sense with regard to irrigation in a riverine environment because, as anyone who has been around them knows, all rivers and their tributaries basically diverge radially from point sources.

The *ceques* that divided the Inca kingdom (as noted earlier, they called it Tahuantinsuyu, or "the four quarters of the world") extended all the way to the ends of the empire and to the limits of the world. Their emanation point was the Coricancha ("Golden Enclosure"), the Inca Temple of the Ancestors. They positioned the great temple squarely at the center of the empire, the junction point of the two rivers. Today, like so many Native American ruins, it is topped by a church—the Church of Santo Domingo—but the remains of a round Inca wall still can be seen beneath it.

The *huacas* (*guacas*) that marked out the *ceques* were natural or human-made temples, intricately carved rock formations, bends in rivers, fields, springs, hills—even trees. In most cases the water theme and its association with the agricultural calendar get heavy emphasis in Cobo's description. The exact location of these *huacas* must have been very important to the Incas, for the chronicler takes the time to describe and pin down all 328 of them. Cobo's descriptions are loaded with concrete information connecting agriculture and the flow of water with what happens in the sky. Inca knowledge about the environment also seems to be closely tied to aspects of everyday life, particularly to the specific assignment of rites of worship and sacrifice.

Where did the people fit in? Cobo tells us that each *ceque* was assigned to one of three hierarchical groups, representing the three social classes (extended families, or *partialities,* as Cobo calls them). Each group was required to tend to their assigned *ceques.* There were, in descending order, the *ceques* that were maintained and worshipped by the primary kin of the Inca ruler, those that were worshipped by his subsidiary kin, and finally those tended to by that segment of the population not related by blood to the ruler. The attendants of these *ceques* came to their *huacas* to offer the appropriate sacrifices to Pachamama at the proper times, Cobo tells us. The assignments on the hierarchy of worship rotated sequentially (from primary to subsidiary to unrelated and back again to primary) and proceeded in quarters from one *ceque* to the next, all the way around the horizon.

The organization of communal work activity, particularly having to do with agriculture and irrigation, was yet another facet of life in the old Inca capital prescribed by the *ceque* map. Representatives of each of forty families drawn from the four quarters of the city participated in a mock or ritual plowing that took place every year in the town square just before planting time. Each delegate would plow a designated portion. Rules for servicing and maintaining the irrigation canals also were specified by the order of the *ceques.* Physically dividing up a central ceremonial place as a field for negotiating social relationships in this way still persists today in remote Andean communities.

Some disagree over whether the actual locations of the *huacas* were preplanned or chosen later to depict the boundaries of the irrigation zones; and not everyone agrees on just how straight the *ceques* really were; indeed, most of them zigzag and a few even cross over one another. What is clear, however, is that the overall plan is basically a radial one (Figure 24).

I became interested in Cusco's *ceque* map because I thought astronomy had something to do with it. I knew the sky was configured into the master plan

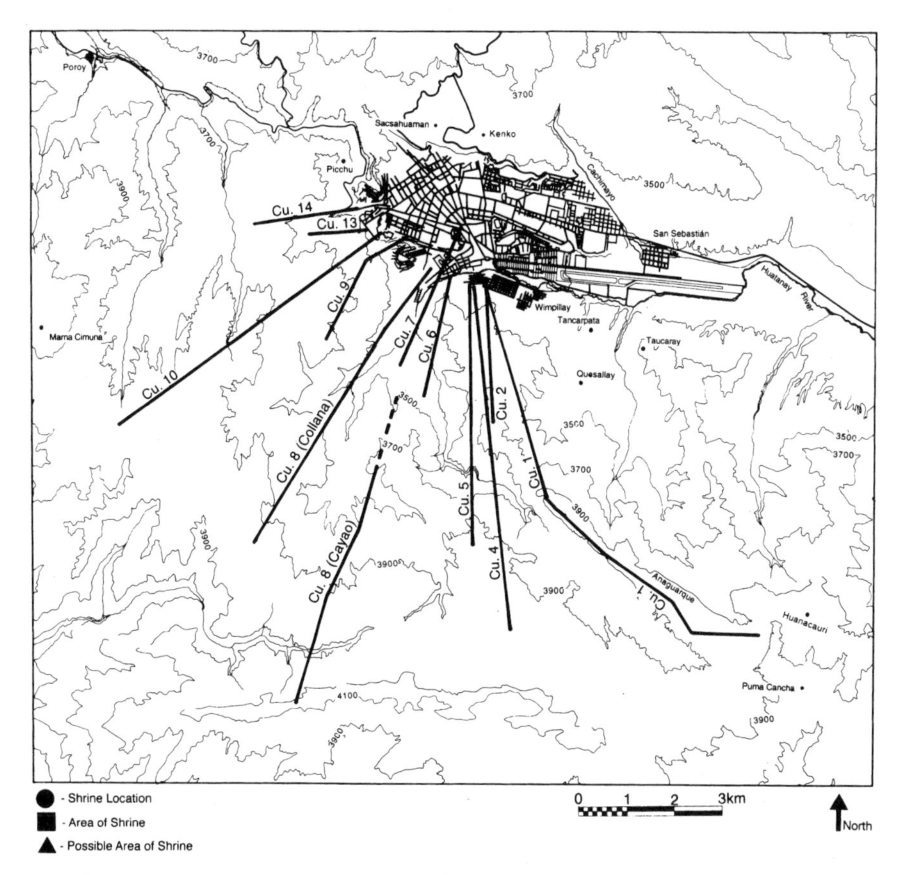

24. Schematic of a portion of the Inca ceque *system, a radial map. (Brian Bauer,* The Sacred Landscape of the Inca, The Cusco Ceque System *[Austin: University of Texas Press, 1998], map 8.3)*

because Cobo's description mentions astronomical *huacas*—sun pillars on the visible mountain horizon that marked important dates in the year. Cobo describes one such set of four pillars when he locates the *huaca sucanca,* named after a hill close to a spring in the northwest quadrant of the city. These overlook Cusco from a headland between two tributaries that descend into the 11,500-foot-high Cusco valley from the northwest, he tells us. This *sucanca* must have been a very important *huaca* because other chroniclers have also written about it. One passage in particular intrigued me because it was so detailed and specific:

> When the sun passed the first pillar they prepared themselves for planting in the higher altitudes, as ripening takes longer. . . . When the sun entered the space between the two pillars in the middle it became the general time to plant in Cusco; this was always the month of August. And when the sun stood fitting in the middle between the two pillars they had another pillar in the middle of the plaza, a pillar of well worked stone about one estado [about six feet] high, called the *ushnu,* from which they viewed it. This was the general time to plant in the valleys of Cusco and surrounding it.[10]

Finding the actual location of *sucanca* in the field seemed a real possibility. Cobo had mentioned several *huacas* both nearby and farther out along adjacent *ceques* flanking the hill. Some bore familiar present-day place names. We even knew the name of the hill—Carmenga (today Cerro Picchu)—and the names of *huacas* along the irrigation canal of Chinchero that enters the valley just north of that hill.

Hill hopping is a necessity if you're going to trace *ceque* lines—which is exactly what my colleague Tom Zuidema, the University of Illinois anthropologist who first wrote about the *ceque* system in the early 1960s, and I intended to do when we began joint studies in the late 1970s. Our method consisted of finding identifiable *huacas* located farthest from Cusco and tracing chains of them along their *ceques* inward toward the city. But finding a source of funds for hill hopping in search of a long lost Inca coordinate system was not so easy. With some effort, we were fortunate enough to acquire sustenance from Earthwatch, a nonprofit organization that matches field projects run by professionals with educated, adventurous tourists desirous of working with them.

To make a map we had set up our surveying equipment on the place where the *ceque* line containing the pillars crossed Carmenga hill. From that point we shot a sight line down to the middle of the plaza. Since the *sucanca* lay on the *ceque* line we were attempting to trace—and since the anonymous chronicler tells us that the sun was sighted from the *ushnu*—we could now fix the important line to the setting sun that crossed this *ceque* and thus calculate the corresponding date in the agricultural calendar. As a bonus, we were also able to locate the site of the *ushnu.* Historians had already known that it had been somewhere in the main plaza of Cusco. The horizon pillars are now long gone, fodder for a later era of builders. The chronicler Garcilaso de la Vega, who mentions having seen them standing in 1560, says the Spaniards had destroyed

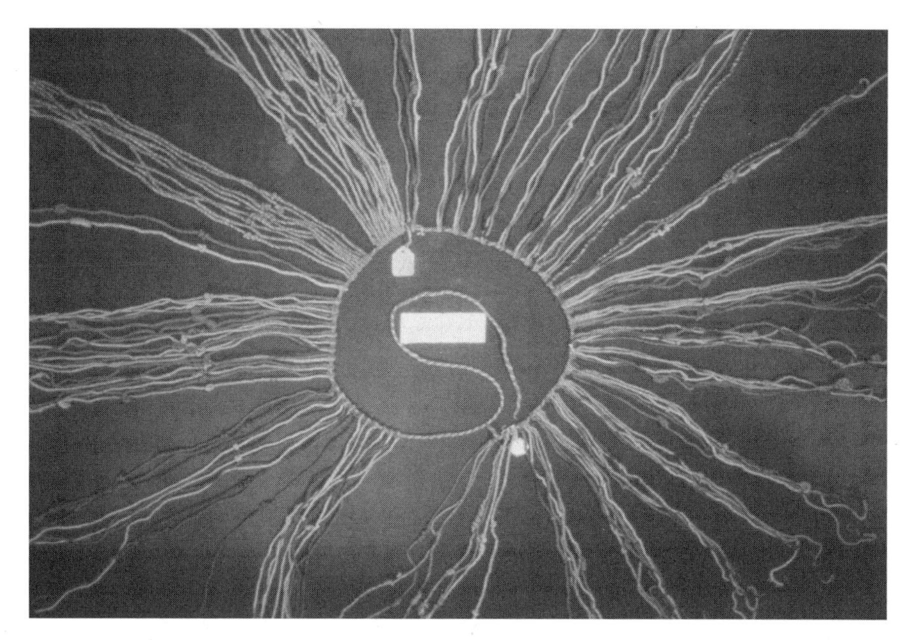

25. The Inca quipu is a linear device made of knotted strings. Does the ceque *map of Cusco mimic its structure? (Museum für Völkerkunde, #VA42593; photo courtesy of Gary Urton)*

them by the time he had grown up out of fear of continued idol worship on the part of their subjects. We erected our own *sucanca* on the spot by piling up stones. Sighting them from the *ushnu* below, we found that sunset at the middle of the pillars corresponded to August 18, the day when the sun passed through the antizenith (one of two days in the year opposite those when the sun passes the overhead point).

Inca astronomers were clever symbolists. They had erected a set of pillars in horizontal space that marked the times of year when they should plant in the descending climate zones in the vertical environment of Cusco. It all centered about mid-August, when the sun crosses the up-down axis and when mother earth is open to being fertilized by the hand plow, whose action would accompany the anticipated rainy season. Brilliant! But is the *ceque* system a real map, I mean more than a mnemonic or mental map? I believe that it is.

Andean *ceques* are reminiscent of Andean *quipus* (Figure 25). The quipu was a tactile form of language, rather like Braille, yet more complete because you can use your eyes as well as your hands to read it. A typical quipu consisted of a thick cotton cord (the primary cord). From it were suspended thinner, secondary cords containing clusters of knots. Often other cords dangled from these, the hierarchy sometimes reaching the fifth order. In many cases the cords exhibited different colors and color combinations and different directions of twisting of the component fiber.

The sixteenth-century chronicler Felipe Guaman Poma de Ayala tells us just how important quipus were. He discusses and illustrates the duties of the *quipukamayoc,* or quipu specialist. Garcilaso de la Vega's later writings on the subject leave no doubt that one of the primary functions of the quipu was accounting:

> They knew a great deal of arithmetic and had an admirable method of counting everything in the Inca's kingdom including all taxes and tributes paid and due, which they did with knots in strings of different colors. They added, subtracted, and multiplied with these knots, and ascertained the dues of each town by dividing grains of maize and pebbles so that their account was accurate.[11]

Garcilaso also tells us that their messages were "written down" in the form of knots in different threads of various colors (red, for example, stood for war). The knots, forms of knotting, colors and color combinations, and the manner of twisting the component fibers—all were probably incorporated into some sort of multidimensional scheme for storing different kinds of information that has yet to be deciphered.

Andean scholars have theorized that Cobo's written description of the *ceque* system may have been derived from information on a quipu: the *ceques* are the pendant cords and the *huacas* the knots on the cords. Although Cobo is a fairly late chronicler of the Inca (1653), good evidence suggests that he acquired his description from an earlier manuscript dated 1559 (in the first generation after Hispanic contact) and written by one Juan Polo de Ondegardo. Polo was employed together with Juan de Matienzo, a lawyer of fairly high standing in the Spanish colony, in the business of negotiating a peace treaty between the last of the Inca rulers and the invading Spaniards. To this end Matienzo and Polo were also involved in locating and rooting out the pagan *huacas,* a necessity for capitulation and conversion of the natives to Spanish Roman Catholicism. To accomplish his goal, Matienzo writes that Polo found it necessary to interview a number of *quipukamayoc* to acquire the information they had written down on their quipus. Among their artifacts there may have been a complex of radiating knotted strips on a quipu that constituted the original "Tahuantinsuyu Atlas of the Inca World," a physical representation of the diverging segments of the sacred landscape of the Inca—their map of the world.

Micronesian stick charts are about as different from what we think of as maps than anyone can imagine. Still, I insist that they are just that, maps—that is, representations of a perceived real world, albeit a world very different from the one we know. By using a stick chart, navigators of the Marshall Islands, tiny coral atolls in the middle of the Pacific Ocean, were able to transform sensory impressions of their watery environment into a concrete model that helped them deal with the most important concern in their culture—navigation.

Stick charts are maps that tell a navigator where and how to proceed on the sea. They are made up of a network of straight and curved reeds bound together to form a flat frame that can be held conveniently in the hand. The straight and curved lines represent an actual course; their intersections, often represented by tiny cowrie shells fastened at various junctions between the reeds, signify the location of various atolls that sparsely dot the course. The navigator can also use the curves and their intersections on a stick chart to help him remember the behavior of the wave patterns that exist in certain places on the ocean surface to help him find his way.

Refraction and reflection of ocean swells by neighboring islands, many of them well out of the navigator's view, produce wave-interference patterns on the ocean surface. These disturbances cause a sailing craft to pitch and roll slightly—so slightly in fact that the disturbances often cannot be seen, but the skilled navigator can feel the tiniest wave vibrations with his body. Equipped with a catalog of such patterns, he lies on his back in the hull and senses them, or he identifies them by the degree of pressure on his upper thigh and testes against the masthead against which he leans. Hardly a scientific navigational technique, but consider that some of these wave interference patterns of oceanic swells are so subtle and minute that they went undetected and mapped until the 1960s, when oceanographers picked them up via satellite photography.

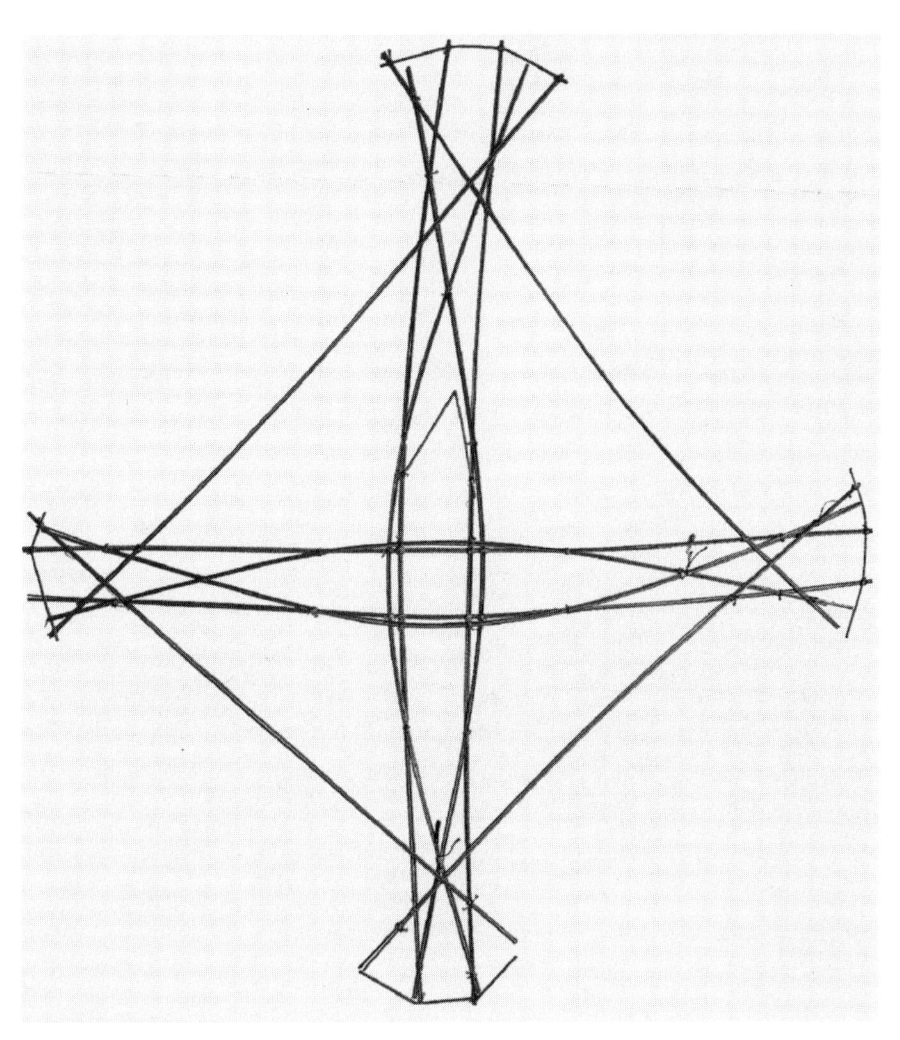

26. This Micronesian stick chart map was once used by a native navigator. (Neg. No. 100920, Field Museum of Natural History, Chicago)

A German sea captain who interviewed a native navigator late in the nineteenth century was the first outsider to learn that these wave patterns served as directional aids. To explain how wave disturbances defined a course, he made a drawing showing what happens when a pair of ocean swells is diffracted by an island. The wave intersections produce a zone of disturbance that increases in strength as the navigator approaches the islands along the line of crossing swells. This line becomes the navigator's course to or from the island

and he must carefully maneuver his craft along the disturbed region, feeling his way as he goes. On the stick chart in Figure 26, the vertical axis represents such a course, and the bent reeds symmetric about the axis signify the swells.

It makes no difference whether the stick chart looks like a realistic map to us. In fact, if you compare one of them with a projection map of the Marshall Islands, it will look quite distorted. But if you think about it, reckoning linear distance at sea is really not a very useful concept in an environment where water, wind, and waves are the stuff the real world is made of. Try driving during rush hour and it won't take you long to sense that the real distance to a place is determined by how long it takes you to get there with or against the traffic or, in the case of the Micronesian navigator, with or against the wind. Motion is the true measure of distance. Some Pacific islanders find it simpler to conceive of their sailing vessel as stationary with the islands slipping by instead of the other way around. (Think about that next time you hang your head over the side of a ship.)

It is important to realize that the two interpretations of the stick chart—as a territorial map and as a scheme for classifying wave disturbance patterns—operate simultaneously. Both serve the purpose of charting the way. For those of us used to a bottom line, this dual meaning is a difficult notion to fathom (and we will encounter a similar case in Chapter 9). That young nineteenth-century German sea captain who viewed the stick chart I just described must have had difficulty, too. When he had trouble learning the mapping scheme system from one of the old native navigators, he was told with some impatience: "You are the dumbest churl I have ever seen, coming every day with the same stupid questions."[12] Only a glass of sherry, which the old man dearly loved, could rekindle his interest in divulging further details of his stick chart system, the eager German interviewer tells us.

You are already familiar (from Chapter 6) with some aspects of the culture from which my final example of somebody else's map comes. It is from the New World and it dates to a time shortly before the voyages of Columbus. When Einstein devised the theory of relativity, he was looking for a format to unite the elusive concepts of space and time. I think the people who created the

27. Time envelops space in this map on page one of the Codex Fejérváry-Mayer from pre-Columbian Mexico. (National Museums, Liverpool, T000627)

Codex Fejérváry Mayer had the same goal. The document is named after the European count in whose hands it landed after most of the native documents were destroyed by Spanish clerics, who considered them impediments to the religious conversion process. It is a painted, accordion-folded document, made out of animal skin, from central Mexico, and dating from about the twelfth century.

Page one of Fejérváry Mayer (Figure 27) is a map of the universe that binds time to space. It depicts a Maltese cross–shaped space in which everything seems to have its proper place. Each quarter of the world, represented by an arm of the cross, contains its associated plants, birds, colors, and directional gods—even parts of the human body are assigned a spatial location. The sun disk rises in the eastern arm of the cross (located, as it is in our European medieval maps, at the top), and death's jaws lie open in the west (bottom), waiting to swallow the dying sun at day's end. As we learned in the last chapter, the people of highland Mexico believed that the sun could be reborn only through human action, specifically, by supplying the nurturing blood of sacrifice. The link between sacrifice and sunrise may be why streams of blood are used to connect the parts of the world with the creator god, who is armed with a spear and atlatl (spear-thrower) at the center of the map. Fejérváry One is also a functioning timepiece. Each of the twenty borders enclosing its cardinal and intercardinal segments contains thirteen circular marks. If we count one circle for each day, we can take a 260-day journey about the periphery of the universe.

We learned about the power of numbers in Chapter 5. Two numbers that lay at the foundation of timekeeping in Mesoamerica are 20 and 260. Probably originating from very ancient times when people counted on their fingers and toes (which is a pretty normal thing to do in tropical climates), 20 was the base of all mathematical systems from the land of the Maya to the northern frontier. We see it carved in stone and painted in manuscripts. And 260? Where and how did this peculiar way of recording time originate? It is unique to ancient America. The Maya call it the *tzolkin,* "the wheel of the days"; the Aztecs refer to it as the *tonalpohualli,* or "count of days." I doubt it is simply the result of multiplying together the two smaller cycles (13 and 20 that make it up) to create a larger one—the way we fold days into months, months into years, and years into centuries, eras, and eons. I think the evolution of 260 is much more complex. It may even emanate from very early times, when the first agriculturists recognized a close correspondence between their own body rhythms and those of the celestial gods they depended on for water, light, and nourishment. (Archaeologists have documented records of using 260 to record time all the way back to 600 BC.) Take the human gestation period and nine cycles of the phases of the moon; Maya women still use it to count their term. Both are nearly 260 days long. Furthermore the female cycle of fertility also approximates the planting cycle in many parts of Mexico. Passages about the Aztec and Maya from Spanish chroniclers tie astronomy to the 260-day cycle. The duration of the appearance of the planet Venus as morning as well as evening star and the cycle of eclipses both jive with the *tonalpohualli.*

At each corner of the Fejérváry map we can read the day name (one of twenty) tied to the first number that begins each consecutive set of thirteen. One Alligator (top right) is followed by One Jaguar (top left), then One Deer, One Flower, One Reed, and so forth. Whereas our weekdays are named for the celestial gods—such as the sun (Sunday), moon (Monday), and planets like Jupiter/Thor (Thursday) and Mars (Tuesday—think of Mardi Gras)—in Mesoamerica they were named for animals and plants, all related to life. Look more closely at the organization of the map on Fejérváry One and you'll find that these same twenty day-name symbols are spatially arranged in four groups of five, each assigned to one of the four cardinal directions; they appear alternating with blood streams in the narrow spaces between the appendages of the cardinal and intercardinal cross segments. Assigning spatial directions to time is something quite unfamiliar to us, especially since we no longer rely on sundials.

Fejérváry One is a goldmine of information on what ancient Mesoamerican people thought about time. If you allow your eye to pass from tip to tip of the intercardinal petals of the cosmogram, making four stops per counterclockwise circuit, you'll reach, at each resting point, the name day of consecutive New Year's days. Thus, the journey through time around this exotic map not only marks out the sacred *tonalpohualli,* but also follows the annual course of the sun—it unites the 260- and 365-day cycles.

Thirteen trips around the world map equal fifty-two years (as well as seventy-three complete cycles of 260 days). Such an interval would have corresponded to an extraordinarily full lifetime of experience for the Aztecs, a "binding of the years," as they called it. One Spanish chronicler quotes an Aztec informant who told him what happens at a binding-of-the-years ceremony: "They put out fires everywhere in the country, cast the statues of their deities into the water, disposed of the hearthstones. . . . [E]verywhere there was much sweeping."[13] The completed round of the calendar signified that time would be renewed—it would repeat itself. Reminds me of our New Year's resolutions, when we cast off our old flaws and resolve to begin anew, marking the beginning of a new cycle of time.

An Aztec map of their own capital city (Figure 28), although made after Spanish contact, has many of the same features as Fejérváry One. It stands in stark contrast with Cortes's map of the city he conquered. Although both

28. This post-conquest map of Tenochtítlan, the Aztec capital, bears a distinct resemblance to the map in the Codex Fejérváry-Mayer made before contact. (Bodleian Library, University of Oxford, Codex Mendoza, M.S. Arch. Seld. A.1, fol. 2r)

display the fourfold structure that organized the Aztec way of thinking discussed in the previous chapter (just as concentric spheres organized Aristotle's model of the universe), the native plan shows the Venice-like nature of this island capital situated in the midst of a lake. Four streams of water merge at the center, which is crowned by an eagle perched on a cactus—still the national symbol of Mexico. As implied in the *ceque* system of Cusco, people are depicted on this map. These figures represent chiefs of the cities located around the lake basin, who paid tribute to the emperor. Like Fejérváry One, the Aztec map is surrounded by time. The names of New Year's Day, written in 260-day notation, that begin each of the years making up the 52-year cycle of renewal run around the outside of the map. Both are maps that represent more than territory.

My last three examples demonstrate that cultures other than our own have created diverse, complex representations of the world. The earthly environment provides many directions for the shape of human thought. We all make maps. To understand them we need to understand the values, beliefs, and motives—and consequently the choices—of the people who made them. But I also focused on these exotic maps to make another point: to know only the innovations of the present and of people like ourselves (the here and the now of our own space-time so to speak) is to live life in a single dimension—life in black and white; on a long, thin wire of time, where one is constrained to move only back and forth. I prefer the less confining three-dimensional world of depth and color that comes with an appreciation of the Other.

As an astronomer, I was interested in studying *ceques* because I saw astronomy in them. Once immersed I learned that the people who built the ancient city of Cusco were using this gigantic overlay map as a way of uniting principles of kinship, geography, dynastic history, rites of survival, calendar, and, yes, astronomy. All are woven together inextricably in the *ceque* system. Tug only on the string of astronomy and the fabric falls apart. Micronesian stick charts seem to be mostly about getting there—although I would wager that there is more to the art of the navigator, an art neither practiced nor remembered there anymore—but they also are maps that tell about the behavior of the ocean, which the navigator must master to achieve the goal of getting there. For the ancient Mexicans, maps like Fejérváry One organized the world and helped forge relations among people as well as between people and nature.

All of these devices for representing the real world exhibit a cosmic and ecological wholeness reminiscent of Homer's model of the world, a quality that we seem to have lost sight of in our modern, highly specialized Western maps. Today I no longer think of maps as mere collections of unrelated elements housed in a large framework. I see them more as sets of properties that, taken by themselves, lose their essence. A map is more than the sum of its parts.

What, then is a map? Better rethink that earlier definition of graphic representations designed to help us understand spatial arrangements. I doubt there can be phrased a good cross-cultural definition that everyone would agree on; but having walked the turf of Aztec Mexico and Inca Peru and sailed the spaces between archipelagos, and having listened attentively to native people who tell us how they see their world, let me give it a try.

To do so I'll elicit the help of historian of cartography, David Woodward. Woodward co-edited a massive six-volume world history of cartography. The project has spanned two decades and is still not complete at this writing. Informed by the richness of non-Western representational devices contributed by experts from around the globe, Woodward goes to great length (several pages) to frame a definition—which he then takes several more pages to criticize.[14] But I think he and I would agree that whether it be organized images in a material medium or a poem, a song, or even a gesture, a map is a way to enhance a people's understanding of place.

NOTES

1. Jorge Luis Borges, "On Exactitude in Science," in *Extraordinary Tales,* ed. Jorge Luis Borges and Adolfo Bioy Casares (New York: Herder and Herder, 1971), 123.

2. Strabo, *The Geography (Geographica)*, tr. Horace L. Jones (London: Loeb Classical Library, 1917–1932), i, I:8–9.

3. Homer, *The Odyssey*, tr. Emile Rieu (New York: Penguin, 1946), 6:95 (verses 271–277).

4. Ibid.

5. Ibid.

6. Ibid.

7. Ibid., 96; cf. *Homer's The Iliad,* ed. and with an introduction by Harold Bloom (New York: Chelsea House, 1987), 22:26–31. The other two winds were Eurus, the east wind, and Notus, the south.

8. Ezekiel 5:5.

9. Bernabe Cobo, *Historia del Nuevo Mundo,* Bibloteca de Autores Españoles, 91–92 (Madrid: Atlas, 1956 [1653]).

10. Anonymous chronicler, Discurso de la Sucesión I Gobierno de los Yngas, ed. Victor Maurtua, in Júicio de Límites Entre el Perú y Bolivia, *Prueba Peruana* 8 (Madrid: Chunchos, 1906), 149–165.

11. Garcilaso de la Vega el Inca, *Royal Commentaries of the Incas and General History of Peru,* 1, tr. Harold V. Livermore (Austin: University of Texas Press, 1966), 124.

12. Captain Winkler, 1901, "On Sea Charts Formerly Used in the Marshall Islands with Notices on the Navigation of the Islanders in General," *Annual Report of the Smithsonian Institution* (Year 1899), 487–508.

13. Bernardino de Sahagun, *Florentine Codex: General History of the Things of New Spain*, 4–5, tr. Charles E. Dibble and Arthur J.O. Anderson, *Monographs of SAR* (Santa Fe: School of American Research and Ogden: University of Utah Press, 1957), 143.

14. John B. Harley and David Woodward, *The History of Cartography,* 2:2 (Chicago: University of Chicago Press, 1984), 210.

EIGHT

LISTS: DOES EVERYBODY DO SCIENCE?

For all observation of the natural world involves the use of mental categories with which we, the observers, classify and order the otherwise incomprehensible mass of phenomena around us; and it is notorious that, once these categories have been learned, it is very difficult for us to see the world in any other way.

—HISTORIAN KEITH THOMAS[1]

There is an outrageously comical but racist (if both are possible) painting from early eighteenth-century Austria titled *Brief Description of Types to be Found in Europe and Their Different Characteristics* (Figure 29). It pictures ten stereotyped gentlemen of varied ethnicity decked out in high-class finery and arrayed in a continuum from the Spaniard on the extreme left to the Turk on the extreme right, with men from the representative nations, including the Frenchman, German, Englishman, and Pole, between. This continuum more or less follows the west to east spread of the nation states on the map of Europe. Beneath the picture appears a list of seventeen archetypal characteristics, arranged

29. *A weird taxonomy:* Brief Description of Types to be Found in Europe and Their Different Characteristics, *eighteenth century, Steiermark, Austria. (From Jack Goody,* The Domestication of the Savage Mind *[Cambridge: Cambridge University Press, 1977], 154–155; reprinted with the permission of Cambridge University Press)*

in a grid so that each personage can be compared with the others regarding manners, personality, intellect, faults, excesses, pastimes, the animals that they resemble, and so forth. The typical Spaniard, for example, is defined as haughty in manner, the Frenchman is frivolous, the Belgian sly. The disloyal Hungarian, the malicious Muscovite, and the fickle Turk hold down the right end of the continuum. At the middle of the scale we find the openhearted German and the proper Englishman. In manners the English are warmhearted, the French (to their left) garrulous, the Poles (to their right) timid. Drinking, working, and eating are the favored pastimes of the central Europeans, but those who inhabit the extreme east and west ends of the continent indulge in intrigue, brawling, and sleeping. I love the animal comparisons: Germans are lions, English are horses, Swiss are oxen; but the French are like foxes and the Muscovites like donkeys.

I have discussed this table with my students on many occasions to demonstrate just how far ethnic stereotyping can be carried. At a more regional level, I remember my Neapolitan relatives describing the neighboring Calabrese to the south as backward; the Marchigiano to the east as stubborn; and the Romans to the north as uncouth (all other Italians despise the Romans). As for the Florentines, Lombards, and other northerners, they weren't even real Italians.

The "Description of European Types" is a taxonomy, a system of classification based on observation and interpretation. As historian Keith Thomas points out in the epigraph, once you learn the system it is difficult to break the hold it has on the way you comprehend the world. This is why, despite all reason to the contrary, we continue to stereotype. Attributing specific negative characteristics to those who differ from you is probably universal. Stereotyping serves as a means of raising your values, customs, attitudes, and appearance above theirs. It helps foster your identity—a positive feeling about being part of your own group. We acquire a false sense of strength and pride by assuring ourselves that we belong to a class or set that manifests superior qualities.

An anthropologist who worked among an isolated mountain people tells the story of the time he asked the headman in his village whether he had ever taken an interest in the people who lived "over there," pointing to the village on the adjacent mountaintop. (Although only a few air miles away, the next door neighbors were nearly two days away by foot.) "Those people are fools," retorted the chief. When time permitted him a week's relief from his ethnographic work, the anthropologist made the two-day hike, returning with a Polaroid shot of the other village's chief. When he confronted the headman with the photo he was promptly lectured: "I told you those people were fools. Look at him—wearing his manta (mantle) over the left shoulder. Even a child knows the manta is worn on the right!"

The human compulsion to find order amidst chaos drives the universal passion to classify just about everything under the sun. We sort out into categories the cars we drive, the clothes we wear, the houses we live in, and the sports teams we cheer for or play on. Americans are especially preoccupied with classifying the food they eat.

The Book of Lists, which topped the American best-seller lists back in 1970s, is a 500-page compendium of lists that ranks, among other things, the ten greatest writers, scientists, and racehorses, along with the five most hated and feared persons in history; the thirty best places to live; the ten fugitives with the longest in tenure on the FBI's Most Wanted list; the ten worst generals; and the ten most beautiful words in the English language.[2] A bit on the fringe (but nonetheless entertaining) are lists of the world's famous hemorrhoid sufferers, people who died during sex or were destroyed by spontaneous combustion, and the most celebrated events that happened in a bathtub!

The earliest taxonomies began as simple lists—inventories of tradable items, catalogs of names of people and the duties assigned them, chronological king lists, lists indicating action or itineraries (like our shopping lists), and lexical lists that ordered items by observable named categories (think of how you'd arrange a roomful of items that belong to an estate you are about to auction off). The earliest scientific taxonomy for which we have written evidence comes from the Mesopotamian ruins of Tell Harmal just outside Baghdad—a 4,000-year-old clay tablet inscribed with the names of hundreds of trees, birds, and other living things, each recognized by their cuneiform signs.

Most of us would tend to associate the business of taxonomy with science, particularly with classifying plants and animals. You really can't do science without it. But why do so many of our accepted scientific taxonomies give us results that seem counterintuitive? A tomato is a fruit, but produce clerks at your supermarket overtly classify it as a vegetable by virtue of where they bin it. Wild rice is closer to grass than to white rice, but you can't find it in the home and garden section. Corn is grass too. Onions and lilies are cousins; bats are more akin to whales than to crows, and whales more closely related to kangaroos than to sea bass. Most popular misclassifications, such as grouping tomatoes with vegetables, stem from not really knowing much about the scientific principles that underlie taxonomies, not to mention a general public unwillingness to observe and name things the way scientists do.

Every branch of science is structured around its pet taxonomies. Name a science and I'll give you a taxonomy tied to it. Geology: I can still remember the three basic rock types—igneous, sedimentary, and metamorphic. Physics: the ROYGBIV spectrum of colors comes to mind. My favorite example from my own discipline, astronomy, is the spectral classification sequence for stars: OBAFGKMRNS, which was designed late in the nineteenth century to characterize the smooth continuum of change in the observed properties of the dark line spectra of stars so that they could be connected to changes in temperature (from hot to cool) and changes of color (from blue to red) along the sequence. The spectral sequence originally started out as ABCDEF, but as technology advanced and observational data increased, spectroscopists realized they had messed up and that some letters needed to be omitted, switched, or combined. (Contorting ABCDEF to OBAFGKMRNS gives you an idea of just how much they had messed up!) All astronomy students are amused by the simple, old (but chauvinistic by today's standards) mnemonic for calling up the sequence: "Oh, Be A Fine Girl, Kiss Me Right Now, Sweetie!" I am among a host of astronomy teachers who will do anything to enliven their classes, including challenging students to update the old list by holding a "spectramnemonics" contest. Here are a few favorite entries:

> "Oh bother! Anthrax found Gary's kitchen! My ramen's now spoiled."
> "On birthdays, Americans fondly give kids many rough, needless spankings."
> "Obstinancy breeds anger, for great karma requires no selfishness."
> "OSX battles antiquated functions. Gates keeps Microsoft, repealing network security."
> "Outside, botching another field goal. Kickers met resistance, no score."

One of the most ingenious taxonomic schemes ever devised for ordering the material world is chemistry's periodic table of the chemical elements. In his book, *Uncle Tungsten,* Oliver Sacks tells of his youthful, geeky fascination with the great periodic table housed in South Kensington's Museum of Science.[3] Each panel in the two-dimensional array displayed an actual sample of the resident element: dark iodine crystallizing in the neck of its bottle; a lump of feisty sodium safely immersed beneath a sea of kerosene; a chunk of glittering, lead-like bismuth; and green chlorine gas translucently visible through its foggy receptacle.

The periodic table was said to have first appeared in a dream to its 1860s creator, the Russian chemist Dimitri Mendeleev, after an insomniac's night of playing solitaire. The primary basis for its organization, as in most scientific taxonomies, resides in the reductive principle—the idea that whatever property of matter we observe is part of a hierarchical branching system that is ultimately traceable (reducible) to a core source. The tree analogy works well here, and housecats serve as a classic example. Think of housecats as residents of the tip of a branch of the tree of life. Follow that branch downward and it merges with other feline branches: lions, tigers, lynxes. Farther down the tree felines and canids merge with a host of other animals (e.g., bears and raccoons) in the carnivore branch, which connects farther down the tree to mammals, ultimately ending in the common trec trunk that represents the animal kingdom. Boxes within boxes is another analogy of hierarchical classification. The all enclosing animal box encases the phylum, class, order, and genus boxes. The species is the "cat box" at the center. Likewise, the periodic table is the trunk or the big box that embraces all matter and the branch tips or centers of boxes are the elements—substances that cannot be further decomposed or reduced to yield other substances—or as eighteenth-century French chemist Antione-

Laurent de Lavoisier put it: "[T]he last point that chemical analysis is capable of reaching."[4]

By the middle of the nineteenth century, experimental chemists had isolated a long list of such elemental substances; but the breakthrough in finding a relationship that would bring them into some kind of systematic sequence came only with the discovery that families of elements—that is, elements having similar combinatorial properties—were dispersed throughout the list. Fluorine, chlorine, bromine, and iodine all combined violently with metals to form white-colored salts, whence the name halogen ("salt former") for this family. The degree of combinatory violence, along with the melting and boiling points of the halogens also changed uniformly with the weight of a given sample. Likewise, silver and copper seemed to behave chemically like gold, while tin and bismuth looked like less ponderous versions of lead. And so, like the game of solitaire, the linear list became a two-dimensional array, with increasing weight running from left to right and the families lining up vertically.

Scientific taxonomists aren't just stamp collectors. Behind every successful taxonomy lies a sound scientific theory and the ability to see through superficial physical differences. The genius of Mendeleev lay in going beyond distinctions such as metal versus nonmetal, to deeper properties like valency, or the relative ability of a given element to combine with other elements according to their atomic weights. By "periodic" Mendeleev meant that in addition to the gradual change in physical and chemical properties within the vertical groups, one could also detect a systematic, repetitive change in the horizontal sequence from the lightest element (hydrogen) to the heaviest (uranium, at the time Mendeleev devised the table).

One criterion for a healthy scientific taxonomy is the preservation of reasoned gaps to accommodate future knowledge. "Vacant places . . . shall be discovered in the course of time," predicted Mendeleev when he laid out his original gap-toothed scheme.[5] So you get the idea: all taxonomies are based on continuous change of observable properties. The closer you look at the details, the more finely honed, all-inclusive, and reductive the taxonomy becomes—if it's a good scientific one.

Unlike many old taxonomies, the periodic table is alive and well a century and a half after it was conceived. Recently two new elements were added to the handful of missing slots in the scrabble board–like arrangement that still adorns science lecture halls. Like the aptly named technetium, these elements are creations of the human hand with a heavy assist from technology. Some of them cannot exist for more than a millisecond because they are unstable, but insofar as their chemical properties are concerned they do indeed fit into their designated slots.

Today's aficionados of chemistry's periodic table are trying to synthesize elements that they believe exist beyond the unstable portion of the heavy end of the table, described by Sacks as "an Alice-in-Wonderland realm where bizarre and gigantic atoms live their strange lives": an island of stability "where the atomic wanderer, after decades of struggle in the sea of instability, might reach a final heaven."[6] I wonder what a "super-gold" or "super-lead" would look like?

So where does all this high-minded classification and list-making leave us? To the average person scientific taxonomies of the material world seem innocuous enough. At least on the surface they have little practical value, although most of us are aware that somehow scientific discoveries are based on their use. The periodic table, for example, led to the recognition of properties of new elements like germanium, which turned out to be seminal in the development of transistors. The recognition of astronomy's spectral sequence helped us to acquire a better understanding of the process of ionization of exotic chemical elements, and geology's firmer foundation for the classification of terrestrial rocks has aided hugely in oil exploration. But ordinary people get more personally involved with the implications of taxonomies when it comes to the principles we employ for classifying living things.

Credit for the invention of the basic taxonomic system that organized the plant and animal world—the genus, species, and specific difference categories—goes to Aristotle. He called them *genos, eidos,* and *diafora,* respectively, and they were taken to apply as well to inanimate things. "For everything," he wrote,

> that differs, differs either in *genos* or in *eidos*; in *genos* if the things do not have their matter in common and are not generated out of each other, as are the beings that have forms different in category; and in *eidos* if they have the same *genos*.[7]

In other words, you can't combine things of different *genos,* which means they can't reproduce. The sub-categories of *eidos* and *diafora* recognize otherness and difference within one another. So, the only difference in *eidos* lies within things of the same *genos*. (*Diafora* has to do with difference of degree.) For example, within the *genos* of serpents, Aristotle recognized an *eidos* division according to the disposition of the internal organs that resulted in some serpents giving birth to live young (ovoviviparous) and other serpents laying eggs (oviparous). Within these species are varieties based on difference in size, diet, and the presence or relative complexity of organs (particularly reproductive organs), or as Aristotle put it, the "degree of perfection" of the product of generation. Above all lies man. At the end of the chain of being, he tells us, man is the perfected form of which lower animals, such as a badly proportioned man, reveal only traces. Surprisingly, it was Aristotle—not Charles Darwin—who first linked the human and animal kingdoms, and it was upon Aristotle's foundation that Darwin would build when he posited the animal ancestry of humans.

Aristotle's system of classification is based on detailed, objective, and dispassionate observation. He presumes that the plant and animal worlds have lives of their own—a far cry from the systems of classification developed in the Roman, medieval, and Renaissance worlds, when plants and animals were categorized not according to sexual proclivity but instead to their relative usefulness to people. One seventeenth-century taxonomy classified herbs as pot herbs, medical herbs, flowering herbs, and weeds. Medical herbs were further subdivided according to which part of the body they healed. These practical, human-centered systems could be quite complex. Animals were broken down into edible vs. inedible, wild vs. tame, and useful vs. useless. Under the useful category, dogs, for example, could be "generous," such as those used in hunting; they could be "rustic," when employed in household tasks of necessity (e.g., watchdogs and sheepdogs); or they could be "degenerate," such as common curs used for turnspits (specifically, a long-bodied dog with crooked legs bred specifically to turn an axle on a treadmill). Under "hunting dogs" lay subcategories based on skills—smelling, spying, and speed—and on the kinds of prey they were most skilled at hunting—beasts or birds. Under each of these sub-categories were listed sub-sub-categories naming the specific type of beast or bird that each dog hunted.

Whether a plant or animal was handsome led to some highly subjective classes. Frogs, for example, were foul and filthy inhabitants of smelly, foggy swamps and bogs. Spiders and snakes were loathsome. The image of the snake has failed for millennia to rise above its shady Garden of Eden reputation. The toad-like bullhead (catfish) was ugly of shape, but Renaissance taxonomists were forced to admit that it tasted pretty good. The changing reputation of types of seafood continues apace. I remember the pre-sashimi era when tuna (at least in the United States) was regarded as a low-class creature, practically on a par with the sardine and suitable only for the cannery. Bottom feeders, like grouper, were scarcely worthy of a place on the menu—a waste of panko bread crumbs—and monkfish, whose taste today is compared with that of lobster, was so base that my grandmother felt the need to soak and wash it in milk several times to remove the impurities.

Unlike the system that would inform Darwin two millennia later, Aristotle's system, like those that flourished during the Renaissance, is static; there is no history and no development as one moves up the ladder of life. Nature's creations are presented as fait accompli—unchanging and destined to last forever. There is also a strong moral component to these human-centered systems of organization that, despite all its progress, science has failed to eradicate. Take the way we project our own morals on the kingdoms of the ants and the bees. We speak of worker and soldier ants who sacrifice their lives for the good of the community, and we impose our notions of political order on the monarchy of the beehive. The fox is still cunning, the weasel bloodthirsty, the pig filthy. Keith Thomas is correct: once we set the categories in place, we have great difficulty breaking their hold on us. (Remember the discussion of altruism among animals in Chapter 3?)

This land is your land; this land is my land. This land was made for you and me. Although this popular verse is intended to convey a feeling of patriotism, it really is rooted in the Bible, which accords man the position of domination over the rest of the living world:

> The fear of you and the dread of you shall be upon every beast of the earth, and upon every fowl of the air, upon all that moveth upon the earth, and upon all the fishes of the sea; into your hand, they are delivered.[8]

Given that sort of moral upbringing, how could one not expect that Western man would create a system for understanding the world that pivoted about himself? Whatever exists in the world was put here for him. The whole idea of studying nature was to master and manage it, to subjugate it for the good of humankind. As Sir Francis Bacon would later put it,

> [A] way must be opened for the human understanding entirely different from that hither to known, . . . in order that the mind may exercise over the nature of things the authority which properly belongs to it.[9]

The ox and the horse were created to haul our loads; trees were put on the earth to supply us with firewood and material for building our shelters; bees are here to nourish us with their honey. Cows, sheep, and pigeons were made to be domesticated. Crossbreeding plants and animals is justified if it leads to the improvement of our well being. And cockfighting and bearbaiting? They are for pure entertainment.

I have a sense that most other indigenous cultures of the world are not so human-centered as we are. The Maya, for example, possessed notions of the cosmos bolstered by a faith that the everyday human world was intimately related to the natural world and that these two worlds functioned in harmony. The universe was a distinct whole, with all parts intricately laced together, each aspect influencing the others. Nature and culture were one. The Maya universe was animate—breathing, teeming, vibrant, and interactive. The Maya talked to the stars and listened to the planets. They commanded and evoked, restrained and constrained, made incantations, pressed their ears to the oracle. They saw themselves as mediators in a great universal discourse. At stake was the battle between fate and free will, between body and soul. Not so in the Western world where man has always been unique.

Only man has a rational soul, said Aristotle, for although below the angels, he is above the beasts. And God's ultimate creation wasn't just good—it was *very* good. Only man possessed speech; only man possessed reason, free will, and moral responsibility. That he was created to dominate is a natural principle. Aristotle wrote:

> [R]uling and being ruled are something necessary and beneficial. . . . Examples are the political rule of intellect over appetite and of male over female and the despotic rule of soul over body and of humans over animals. So any human beings related to others as body to soul or as animal to human are slaves by nature and such are those whose best work is the use of the body.[10]

Classifying plants and animals on the basis of their structural characteristics rather than their utilitarian, esthetic, or moral status—that is, *scientifically*—didn't return in earnest until the late seventeenth- and early eighteenth-century Enlightenment, when the notion that classifying was part of the mental process of finding the one true order that corresponded to the structure of the sensate universe—the means to Bacon's end. The idea was to identify every imaginable set of objectively based observational principles that could be filed away into boxes or branches.

Called by one of his many contemporary admirers "God's registrar," Karl Linné—or in Latin Carolus Linnaeus (1707–1778)—was a Swedish doctor-turned-teacher who acquired a passion for collecting lots of things: rocks, stamps, and especially plants. No mere collector, Linnaeus saw as a divine mission the creation of an orderly inventory of all living things. He was among the first to pick up on Aristotle's way of classifying matter—the familiar animal, vegetable, and mineral categories. Linnaeus wrote in Observation 15 of his masterwork, *Systema Naturae,* "Stones grow; plants grow and live; animals grow, live, and feel."[11] His search for the botanical "thread of Ariadne"—which led Perseus out of the labyrinth after he slew the minotaur and without which there is chaos—answered a question that went far beyond Aristotle's simple questions: what does it look like and where does it come from? Linnaeus's system was the first true biological taxonomy based on sex. Linnaeus touched on Aristotle's what-does-it-look-like question when he fixed the genera of the plant world by observing the varieties of their "organs for fruiting." The species were determined by variations in each of the recognized arrangements that made up the genera.

More specifically, the botanical system of Linnaeus is based on the size, number, and location of the stamens (the male organ, consisting of a number of filaments) and the pistil (the surrounding seed-bearing organ). The same holds true in the animal world. Where individual species of butterflies and bees belong boils down to the fine structure of their genitalia. Closer to the trunk of the evolutionary tree, the quadruped genus differs from the bird genus by virtue of the former being ovoviviparous, whereas the latter is oviparous. The system was hierarchical and dichotomous, with the detail of information increasing as one moved down the hierarchy. For example, all shrub-borne currants belong to the genus *Ribes* (Latin for "currant"); thus, all species within this genus bear the name *Ribes.* To complete the name, the specific species name, often quite descriptive, is added to the genus name; the species of red currant is referred to as *Ribes rubrum,* that of the black currant *Ribes nigram,* the flowering currant *Ribes aureum,* and so on. This system has been adapted to all living forms, often with far less logical names. Given the myriad species of recognized insects, I suppose that it was inevitable that contemporary binomial taxonomic notation would begin to exhaust all possibility. Out of boredom entomologists have turned to naming species after celebrities (*Orson-welles malus*—a type of spider), famous quotations (*Ba humbugi*—a type of snail), puns (*Pieza pi, Pieza rhea,* and *Pieza deresistans*), and palindromes (*Xela alex*). One group of taxonomists has recently offered to auction off naming rights to a newly discovered species of monkey.

Some of Linnaeus's contemporaries were shocked by the idea that sexual characteristics could matter in the order of creation. "[S]uch loathsome harlotry as several males to one female would never have been permitted in the vegetable kingdom by the Creator,"[12] wrote one detractor. And there were others. Remember the *Brief Description of Types* painting? Nationalism made it difficult for British taxonomists to deal with a superior system conjured up by a Swede. John Ray, their favorite son, had also examined the detailed structure of plants; his system asked, were they with or without flowers? Did they have one- or two-sectioned leaves? Were these leaves attached or unattached? and so forth. In a diatribe delivered against traitorous disciples of Linnaeus residing in the British Isles, taxonomist John Fleming wrote that those who followed Linnaeus had "suffered for their folly, in preferring the naturalist of Sweden to Ray, who was vastly his superior in philosophical attainment, enlarged views . . . and good taste."[13]

Despite its detractors, the Linnaean system would survive and lead to the development of several "ologies"—such as zoology and physiology—that would proceed far beyond static tabular arrangements toward the higher intellectual goal of discovering the true history of all living things and man's place in that history. Recently, however, a small group of rebellious taxonomists has sought to overturn the system.[14]

As scientific taxonomies took over the natural world, the old human-centered systems of the Renaissance were relegated to the realm of occult superstition to be poked fun at rather than understood as harmless antiquated traditions. Obviously the principle of creating superior self-identity works over time as well as distance. The gap between scientific and vulgar nomenclatures widened to a canyon. Adding Latin names, one for the genus and the other for the species, only made things worse. Why classify plants by making them unintelligible, so that only a botanist can know them? complained one nineteenth-century lover of nature. If you are ignorant of Latin, then you shouldn't study botany, came the reply. Still, we persist in picking brown-eyed Susans, eating black-eyed peas, and taking St. John's Wort (in modest doses) for what ails us.

The use of vulgar herbal remedies for tuberculosis, cancer, backache, and underarm odor was vigorously attacked too. Likewise, the popular belief in mythical animals was systematically debunked. We may no longer believe in unicorns and mermaids (the latter probably stemmed from a hasty view of a seal, argued one scientific skeptic), but the Loch Ness monster and the abominable snowman (or the "abdominal snowman," as some of my neighboring Adirondack mountain locals who spend long winter eves in snowbound taverns discussing sightings tend to call him) live on. A skeptical movement, not unlike the one that flourishes today, developed out of the great schism between popular and learned scientific ways of classifying the natural world.

Today, the Committee on the Scientific Investigation of Claims of the Paranormal (CSICOP) launches regular sustained attacks on UFO sightings, flat-earthers, channeling, and ESP through its publication, *The Skeptical Inquirer.* The contemporary health-food store, the FDA, and the drug companies are far from being in accord on how to designate what belongs in which box. Plants and animals may be subjects of contemplation by poets, artists, and other romantics, but as far as science is concerned all things have a life of their own. When we study them, we peer through a window into the lives they live for their own sake—not for ours.

Is Western science unique in its quest to order the environment by creating taxonomies based on dispassionate, objective observation? Do other cultures of the world create their own "folk taxonomies" solely for utilitarian purposes—the way our ancestors did before the Enlightenment? Are they like us? If not, what characteristics do these other systems share with our own and what can we learn about the connection between people of diverse cultures and nature by studying them?

What places things in their categories in many folk taxonomies has less to do with structural details about *what it looks like* and more to do with other kinds of detail, such as *what it does.* The Navajo classify by name all living things according to whether they can or cannot speak. They subdivide animals by whether they run, fly, or crawl and they divide them further according to whether they travel on land or in water, by day or by night, and so on down the hierarchy. The Hopi go further. They assign each plant and animal a direction in space. In fact, all directions (the Hopi recognize six: northwest, southwest, southeast, northeast, up, and down) have plants, animals, and colors associated with them. Snakes belong to the nadir (below), vultures to the zenith (above), sagebrush to the southwest, gray rabbit brush to the northeast. This system of directional assignments is reminiscent of what we found in ancient Mexico in the last chapter. These systems do not isolate the plant and animal worlds into so many separate kingdoms; the categories are more interactive, more dynamic.

You may be surprised to learn that ethnobotanists agree that some folk taxonomies are hierarchic and binomial, just like the one devised by Linnaeus. The Tzeltal Maya recognize four classes of plants: *te* (tree), *ak* (grass), *ak'* (vine), and *wamal* (herb).[15] That covers about 80 percent of all plants; cactus, bamboo, and agave are exceptions. There are twenty-eight kinds of *te,* among them the *hihte* (oak). In the next step down the ladder are the five kinds of *hihte: cikinib, sakyok, kewes hihte, capal hihte,* and *cis hihte.* Note that only three of the five subcategories of plants formally carry the word for oak, which means that the name of the thing is not sufficient to place it in a particular class of tree, although Tzeltal informants say that this is where it belongs.

When anthropologists study taxonomies in detail, they find that the connection between language and the environment often holds the key to understanding the system. To demonstrate, let me draw on the outstanding work of anthropologist Werner Wilbert, who has spent much of his life among the Warao, an isolated people of the Orinoco delta in Venezuela.[16] The Warao, who spend much of their lives on the floodplain, shocked Wilbert by recounting an incredibly detailed and precise scheme for organizing knowledge of their environment expressed through their own binomial taxonomic systems. A nonliterate people, the Warao are living disproof of the hypothesis that you need to know how to write in order to create a detailed taxonomy. They recognize three different kinds of sediment: sandy soil (*waha ahobahi*), silty soil (*hobahi ahobahi*), and clayey soil (*hokoroko ahobahi*). They sub-classify each of these according to texture, structure, color, and odor. Under color, for example, they speak of white or light (*hoko*), brown or dark (*ana*), and yellow-orangish (*simo*) soils. Thus, white sand would be *waha hoko* and black clay *hokoroko ana.*

Sharing the reductive strategy of Western science, the Warao further classify the granularity of half a dozen types of silt based principally on how it feels on the bottom of their feet. There's firm silt (*hobahi taera*), semi-firm silt (*hobahi taera sabuka*), orange-like silt (*hobahi simo*), semi-firm water-saturated silt (*hoboto taera sabuka*), water-saturated black silt (*hobahi hobotoboto*). To check the Warao on this point, Wilbert carefully measured the percentage of grains of different sizes that made up these silts. He found a smooth variation across the six types of soil recognized by the natives, ranging between 97 percent and 75 percent of the total comprising grains no larger than 0.083 millimeter in size—pretty remarkable, especially when we realize that these categories are sensed simply by walking barefoot. Some may also see it as remarkable that scientists feel the urge to check the Warao on this point.

Given the watery nature of the environment (with more than a dozen tributaries flowing across the flat delta), it isn't surprising that the Warao also have developed a complex taxonomy of water. They classify seven kinds of bodies of water and they distinguish them by whether they have currents and what types of currents flow in them, whether they are tidal or fresh, and on the nature of their tributaries, sub-tributaries, and meanders; whether their borders are trough-like or channel-like also matters. Western taxonomies lack words for many of the riparian features recognized by the Warao. On the other hand, these people have no word for rock or mountain. That's because, of course, their world doesn't have any rocks or mountains.

Lush with vegetation, the swampy savanna environment has given rise to a Warao plant taxonomy as detailed as that of the Tzeltal Maya. The Warao differentiate between herb plants that have no woody parts (*bebe arau*), vine plants that have long thin woody trunks (*ero arau; ero* means "long"), tree plants that grow without need of support from other plants (*dau arau*), and palm plants (*yawara arau*). They name parts of plants by corresponding parts of the human body. Trees, for example, have heads (crowns) and bodies (trunks).

Leaves have flesh, a chest (the front surface), and a back (the rear surface); their serrated edges are teeth. Leaves also have veins, hair, bones, and they bleed (sap). Fruits have eyes (seeds), flesh, and skin. Despite Linnaeus, those of us who inherited the scientific taxonomy of the West also continue to endow our potatoes with eyes, our leaves with veins, and our trees with trunks.

Amidst this seeming minutia lies the driving force of this curious eco-taxonomy—sustenance. By classifying their environment in such exquisite detail, the Warao have managed to finely tune themselves to the interdependent nature of its components and to the detection of subtle signals that monitor even the slightest changes in a precariously balanced habitat. They interpret these data by telling "transformation stories," or tales that express the relationships among the complex taxa they have created. Let me share just one of Wilbert's many examples, the story of Red and White Cedar.

Red Cedar began her journey in the upper delta; she became a female canoe and traveled to Trinidad. On her return journey up the delta she was accompanied by her younger sister, White Cedar (Canoe Woman). This sister decided to settle on the soggy soil of the lower delta, but the older sister, who didn't like the boggy smell there, transformed herself into an anaconda snake and made her way back up the delta by ingesting lumps of clay. Out of these lumps of clay she created many different shapes of levees and ridges where she could pull up and rest. Finally she reached her place of origin and there she built her home and became the Mother of the Forest.

Today red cedars have taken up root all the way down the delta. The people say they are the daughters of the Mother of the Forest. When they are transformed into canoes, they become the wives of the boat makers and the cedars instruct them in their craft. Far from an isolated item of vegetation with a Latin name tag, the red cedar tree is one element in an interactive ecological system that includes soils, different types of habitat, other plants, and people. It is part of a binomial taxonomy not unlike that of Linnaeus, but it also integrates social, economic, technological, religious, and scientific ideas.

Anthropologists who enter the land of the Warao without much understanding of its ecology will not have the slightest chance of unearthing the subtleties of the types and subtypes of deltaic soils and trees recognized by the local inhabitants, for these highly observant fishermen and foragers name categories in their soil and plant taxonomies that our system of classification cannot accommodate and for which our language lacks words. So don't expect to elicit much information about their native taxonomies by taking bits and pieces of your own and asking, what do you call this? Or that? Asking a Warao, what do you call a rock? would surely elicit a blank expression.

Classifying color is another example where it is all too easy for outsiders to miss the boat. We are all familiar with the ROYGBIV mnemonic of our color spectrum: red, orange, yellow, green, blue, indigo, violet. And most of us have heard of the Munsell color wheel, which invokes the principal combinatory properties of these recognizable hues. Now, suppose a hypothetical anthropologist wanted to investigate the color classification system of some exotic culture. The naïve inquisitor might pull out some Munsell color chips and ask the informant, what do you call this hue and how would you describe its brightness or saturation? These are all standard terms for discriminating among colors in our culture. But that wouldn't work with the Maya of Yucatán, who recognize only five hues or colors: black (*ek*); red, pink, orange, or rust (*cak*); yellow (*kan*); white (*sak*); and green (*yax*); however, these terms appear in seventy-five compound words that discriminate among brightness, saturation, discreteness, opacity, and texture.[17]

Because space forces me to spare you all the linguistic details, I will briefly explore just one element, that of color texture—which has no analog in our culture. The Yucatec Maya language combines the root word for soft or tender with color terms to give compound words that mean, for example, tender black (tender tips of young leaves); or tender red (the skin of a young person affected by the sun); shiny yellowish (nicotine on the fingers); puffy black (a black eye); prickly green (the fruit of a fine spiny cactus); or raw red (your nose from a cold).[18] In the last example the question, why is your nose raw red? takes on an added dimension. It asks not only why is it red, but also why is it rubbed raw. This sort of foreign color taxonomy helps us realize not only how visually oriented our culture is but also how impoverished is its taxonomy when it comes to offering concise, *vivid* descriptions. The Maya, on the other hand, have developed a descriptive taxonomy for specifying detail by appealing to sensory organs other than the eye. If you work with fabric, you may recognize similar textural distinctions, like silky green vs. velvety green, terms that refer as much to how the fabric feels in the hand as to how it appears to the eye.

Color is generally regarded as a superficial quality in Western descriptions of nature, but it emerges in other non-Western cultures as an elemental diagnostic category. Take Inca metallurgy. Plasticity is the property of metal that most interested the Andean metallurgist. An alloy is really a medium of transformation. When an Inca metalworker hammered out a silver-copper alloy, he was said to be engaged in bringing out "the truth of the metal" from its interior.[19] As he hammered, he watched the surface of a sheet of metal gradually change color from red to white—from copper to silver. It gets toughened. The process of hammering brings out the essence of the material, whose true potential lies hidden beneath its surface. The creative process flows along the third dimension—the dimension of depth.

In Andean weaving and Andean astronomy, what happens below also has a direct influence on the outcome, or what is observed at the surface, whether you are working with the surface of a textile or the surface of the earth. As we learned in Chapter 7, in Andean astronomy the vertical duality of the cosmos suggests that what you cannot see (such as the sun at the antizenith, directly beneath your feet) brings out what you ultimately will experience on earth, because when the sun goes there, the rains that signal the start of the planting season will soon arrive. Once it passes beyond the horizon through the midpoint of the underworld, the sun comes back to initiate another turn of the seasons. In these technologies the message comes from the way in which the object or the world is worked, whether by god or by craftsperson.

We are already aware that without taxonomies science could not exist. As we've seen in the first part of this chapter, the sequential arrangement of things and phenomena based on the recognition of a continuum of observed characteristics constitutes the first creative act in the inductive process. And although I haven't run the gamut of the disciplines, if you do so on your own, you will find that basic taxonomic schemes underlie all of the organized "ologies" that make up the physical and biological sciences.

The observed properties of the world around us that beg attention may vary from culture to culture and from environment to environment, but the need to arrange and to organize those properties constitutes, for all of us, the first step toward making nature knowable. Arranging things in hierarchical patterns seems to be as universal as creating compound words to decide categories in which to place them. On the surface the criteria for classification seem to be mainly functional, and the function differs widely from one cultural and environmental context to another. But sometimes the detail that emerges convinces me that when it comes to the business of classification, most of us choose to go beyond mere necessity.

Classification is an art form motivated by curiosity and driven by the need to know where things belong and how to differentiate one thing from another. The criteria for what constitutes order, however, vary widely from culture to culture. All of these taxonomies are deeply rooted in the values of the cultures that gave birth to them. In the biological taxonomies of the West, for example, the sex-based criterion yields recognizable change in appearance between ancestral and derived states, which has its modern foundations in evolutionary theory.

Today few historians would regard science as anything other than a unique phenomenon peculiar to the West—a way of knowing that grew out of a particular set of ideological, socially derived turns taken by the culture that devised it. These turns are based on four key concepts: abstraction, deduction, theory, and observation—the latter usually conceived in the form of experiment. None of these materialize and connect in quite the same way outside the Western sphere, nor do they make up a significant portion of native vocabularies. But seeking them elsewhere is a bit like using our Munsell color chips to interrogate a Yucatec Maya person. All you get back in turn is a vague recognition (or lack thereof) of *your* own concepts in *their* culture. You learn little about *them.*

Already aware that a multitude of indigenous cultures have long been systematizing knowledge for its own sake (which is basically what we're trying to do with our scientific approach), based on what I've learned by studying taxonomies I'd rather alter my definition of what science is from a standard widely accepted version penned by a well regarded historian of science, Alistair Crombie:

> [The search] for the intelligible impersonal permanence underlying the world of change [devised by the Greeks by] hitting upon the brilliant idea of a generalized use of scientific theory; they proposed the idea of assuming a permanent, uniform, abstract order from which the changing world of observation could be deduced.[20]

And move to this more open-minded statement by another equally respected historian of science, David Pingree:

> [S]cience is a systematic explanation of perceived or imaginary phenomena, or else is based on such an explanation.[21]

Clearly, observation is the key. But any mention of the phrase "the observation of nature" ultimately leads to the question of efficacy, which in turn leads to the standard criticism of all ethno-taxonomies leveled by the (usually uninformed) Western practitioner: "But our system works and theirs doesn't." The anthropologist usually counters this declarative statement by citing examples or case studies that, when examined from a point of view centered in the culture, yield a high degree of efficacy (e.g., Warao soil taxonomy). But, "Did these people *really* know what they were doing?" is the usual response. The compromise often reached is that although "They may not really have known what they were doing," trial-and-error experience, together with good luck, evidently produced (sometimes) a cure, an explanation, a procedure that works. But who says validation can only be arrived at by statistics and probability based on data obtained in controlled tests, as is the case in Western science? As one of my colleagues remarked at a seminar on ethnoscience a few years back, "Asking these kinds of questions assumes these people were operating without systematic knowledge."

The Iakoute of Siberia say that contact with a woodpecker's beak is a cure for toothache. This is a statement that someone who grew up under the umbrella of Western medicine would find difficult to take seriously. But as the dean of structural anthropologists, Claude Lévi-Strauss, once wrote:

> The real question is not whether the touch of a woodpecker's beak does in fact cure toothache. It is rather whether there is a point of view from which a woodpecker's beak and a man's tooth can be seen as "going together" . . . , and whether some initial order can be introduced into the universe by means of these groupings.[22]

Making assumptions about how things go together based on our own experience often can deprive us of understanding what indigenous taxonomies really mean. Maybe we shouldn't place so much emphasis on searching out whether a particular procedure works absolutely or objectively. Better we should ask questions that get at the root of *their* systematic explanations (to use Pingree's terminology). What did they think they were doing and how did they know it? And there is the related question that holds the key to our understanding of their culture and its conception of nature, how can we know what they thought they were doing?

Exploring someone else's way of organizing the world can be a rich, expansive learning experience. Personally, I thrill in the challenge of attempting to acquire another conception of the order of things rather than being content simply to retreat into my own. And because I straddle the line—or, better, walk the tightrope—between the scientific and the humanistic disciplines, I pay the price of the doubter. I will never know for sure whether the order I conceive lies in myself or in what I classify. But then, even the late Stephen Jay Gould, a lifelong professional taxonomist, has opined that as long as taxonomies express both concepts and percepts, they must "teach us as much about ourselves and our mental modes as about the structure of external nature."[23]

NOTES

1. Keith Thomas, *Man and the Natural World* (New York: Pantheon, 1983), 52.

2. David Wallechinksy, Irving Wallace, and Amy Wallace, *The Book of Lists* (New York: William Morrow, 1977).

3. Oliver Sacks, *Uncle Tungsten: Memories of a Chemical Boyhood* (New York: Knopf, 2001).

4. Gerald Holton and Duane Roller, *Foundations of Modern Physical Science* (Reading, MA: Addison-Wesley, 1958), 377.

5. Ibid., 419.

6. Sacks, *Uncle Tungsten,* 316.

7. Aristotle, *Metaphysics* ι.1054b27.

8. Genesis 9:2–3.

9. Jerry Weinberger, ed., *New Atlantis and the Great Instauration* (Wheeling, IL: Harlan Davidson, 1980), 17 (preface to *The Great Instauration*).

10. Aristotle, *The Politics,* tr. Ernest Barker (Oxford: Oxford University Press, 1946), 1, V:5.

11. Heinz Goerke, *Linnaeus Systema Naturae* (New York: Scribners, 1973 [1735]), 84.

12. Ibid., 108.

13. Harriet Ritvo, *The Platypus and the Mermaid and Other Figments of the Classifying Imagination* (Cambridge: Harvard University Press, 1997), 25.

14. To accord more harmoniously with evolution, they have proposed devising a PhyloCode that places all living things according to descent from the most recent common ancestor, pointing to the tips of a pair of twigs on the tree of life, extending to where the branches meet. The taxonomic group conforms to everything in between. See Joshua Foer, "Pushing PhyloCode," *Discover* 26 (April 2005): 47–51.

15. Brent Berlin, Dennis Breedlove, and Peter Raven, *Principles of Tzeltal Plant Classification* (New York: Academic Press, 1974).

16. Werner Wilbert, "Conceptos Etnoecológicas Warao," *Scientia Guaianae*, 5 (1995): 335–370.

17. Victoria Bricker, "Color and Texture in the Maya Language of Yucatan," *Anthropological Linguistics* 41:3 (1999): 283–307. I am indebted to Victoria Bricker and to William Baleé, both of Tulane University, for a number of informative discussions of native folk taxonomies when I served as Mellon Professor of the Humanities in spring 2002.

18. Ibid., 288.

19. Heather Lechtman, "La Metallurgia Andina" (Andean Metallurgy), in *Storia della Scienza* 2 (Rome: Marchesi, 2001), 1017–1018.

20. Alistair C. Crombie, *Medieval and Early Modern Science* 1 (New York: Doubleday, 1959), 5.

21. David Pingree, "Hellenophilia and the History of Science," *Isis* 83:4 (1992): 554.

22. Claude Lévi-Strauss, *The Raw and the Cooked* (New York: Harper and Row, 1969), 9.

23. Stephen Jay Gould, *I Have Landed: The End of a Beginning on Natural History* (New York: Harmony, 2002), 303.

N I N E

SYMBOLS: THE NATURAL HISTORY OF A ⊕

. . . [a] peculiar petroglyph carved into basal rock on the southeast slope of Cerro Colorado, a small hill about 3 km to the west of the Street of the Dead. The petroglyph consisted of two concentric circles and a cross pecked into the rock. An identical design [was] pecked into the concrete floor of a room on the east side of the Street of the Dead. . . . The interesting thing about [the] two identical marks is that a line through them forms a nearly perfect right angle . . . with the center line of the Street of the Dead.

—ARCHAEOLOGIST JAMES DOW (1967)[1]

[There is] a great slab of more than 28 square varas [about 16 square yards] on which are marked two circles and in the middle one diameter, then another at right angles to it, all indicated with groups or points giving the chronological periods.

—NINETEENTH-CENTURY HISTORIAN AND WRITER
ALFREDO CHAVERO (1889)[2]

The design, made by small circular depressions in the floor, consists of two concentric circles divided into quadrants by two straight lines. The extremes of the straight lines point to the cardinal points. . . . [T]he straight lines are arranged so that each has ten dots from the center to the inner circle, four dots from the inner circle to the outer, and four dots beyond the outer circle.

—ARCHAEOLOGIST LEDYARD SMITH (1950)[3]

Here are three descriptions of identical ancient artifacts situated thousands of miles apart and spanning almost a century, but I did not hear about them in this order. (These along with other examples are shown in Figures 30–34.) The order mattered, for as I would learn, how we arrive at the truth depends not only on the facts of the case but also on when we get the information.

30. The ubiquitous pecked cross-circle symbol (⊕). This one is pecked into a floor in the Ciudadela in southern Teotihuacán. (Photo by author)

Teotihuácan was the greatest of all the ancient cities of the New World. The fifteenth-century Aztecs say that it was the home of their gods, the place where time began. Built in the third century BC on more modest settlements, the site is about an hour and a half (in moderate traffic) northeast of Mexico City. A major tourist attraction today, it comprises more than 2,000 stone and stucco structures arranged in a magnificent grid stretching over eight square miles of a valley ringed by high mountains. Teotihuácan was once an international city, replete with apartment complexes arranged in ethnic "barrios" that could be identified by the different styles of ceramic remains archaeologists continue to find in various quarters of the city. They estimate that the city's population may have exceeded 100,000 at the time of its rather sudden and inexplicable demise (around AD 600), which was accompanied by a vast conflagration. The great Pyramid of the Sun—which at 240 feet across at its base is as wide as Egypt's largest pyramid and about two thirds as high—flanks the perfectly straight Street of the Dead. It paves the way up to the Pyramid of the

Moon, which lies at its northern terminus, so situated to imitate the shape of the great mountain Cerro Gordo, beneath whose slope it resides. All these names, by the way, are later contrivances. No one really knows what the people of Teotihuácan called their city, much less what they named their buildings.

My first encounter with the pecked cross-circle symbol—let me call it a "⊕" for the sake of space—coincided with my first visit to Teotihuácan's pyramids. It was 1970 and recovering lost knowledge from an enlightened past was part of the mystique that accompanied the 1960s' anticipation of the Age of Aquarius. Stonehenge had just been "decoded" and archaeologists were having a rough time accepting the findings of engineers and astronomers who had unleashed their surveying instruments and magnetic card (programmable) calculators on the Salisbury plain seeking to mathematize Great Britain's most ancient monument. The data they collected on the megalithic alignments and on the spacing of elements that make up Stonehenge's circular design proved (at least to the satisfaction of many) that Stonehenge was not simply a place where barbarian tribes assembled for religious worship. It was also astronomically oriented with a level of sophistication that surpassed any contemporary device in ancient Egypt or Sumeria for gleaning knowledge of the sky.

Somewhere in the fine print of the flurry of reportage on Bronze Age megalithic science appeared the suggestion that perhaps the pyramids of Mexico were also aligned with the cosmos. Since few had ever bothered to explore that question in any detail (and since learning through direct experience was a major outgrowth of the campus revolution of the 1960s), I soon found myself in Mexico, accompanied by a small group of students, standing on the floor of a ruined temple in the 2,000-year-old ancient city of Teotihuácan. We were all looking down at the curious design on the floor described by architect James Dow (in my first epigraph), who had helped excavate it. It was about two feet across and made up of half-inch holes neatly hammered into the stucco floor arranged to portray two concentric circles centered on a cross. It looked like a miniature target for a parachute jumper.

Archaeologist Rene Millon, head of the Teotihuácan Mapping Project, had taken me to see the ⊕ (Figure 31) when he learned that I was an astronomer with an interest in building orientations. Millon was friendly and forthcoming—a lot more so than the British archaeologists who confronted the astronomers who were invading their turf. Later we joined one of his graduate

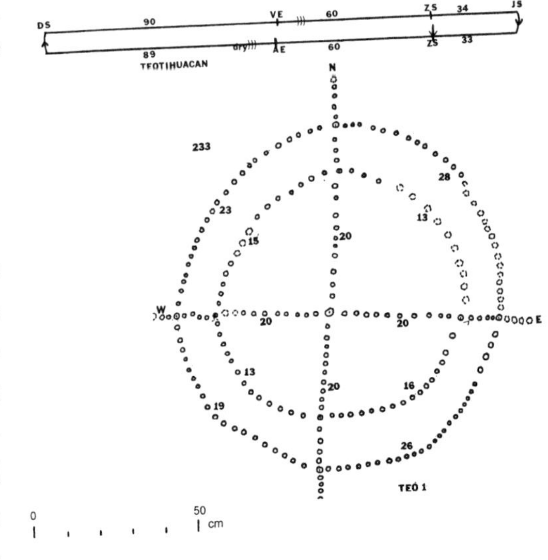

31. The ⊕ hammered into the floor of a building on the east side of Teotihuácan's Street of the Dead. (Drawing by author)

students for a dusty jeep ride to a second design—the one on the flank of Cerro Colorado. Carved on an isolated flat piece of rock outcrop (and therefore a true *petro*glyph), it looked pretty much the same as the one located in the city center (Figure 32). We set up our surveyor's transit (the same kind you see on road crews) over it and sighted a striped pole held by one of my students over the first ⊕ back at the archaeological site some two miles to the east. We worked the measurements through our avant-garde portable, magnetic-card calculators, and—sure enough—the alignment lay exactly perpendicular to the Street of the Dead, the two-mile-long north-south axis of the city. The situation was exactly as Dow had described it.

"Teotihuácan was clearly subject to planning," Millon once wrote, and that's a bit of an understatement.[4] The metropolis was disoriented from the lay of the land, twisted 15.5° off the cardinal directions. Ancient city planners even took the trouble to canalize the northeast-to-southwest-running San Juan River that runs through it, forcing its waters to pass precisely along the skewed east-west grid.

From a practical standpoint, adhering to the torqued Teotihuácan orientation would have necessitated considerably more labor than if one simply followed the contours of the landscape, but the builders of Teotihuácan seem to have chosen this format because of some overriding cosmological or geometrical consideration—an irrational principle, as Millon characterized it. The purpose of the two ⊕ symbols, then, was to serve as the architect's benchmarks, part of a scheme for precisely laying out the east-west perpendicular to the skewed north-south axis to set up the grid.

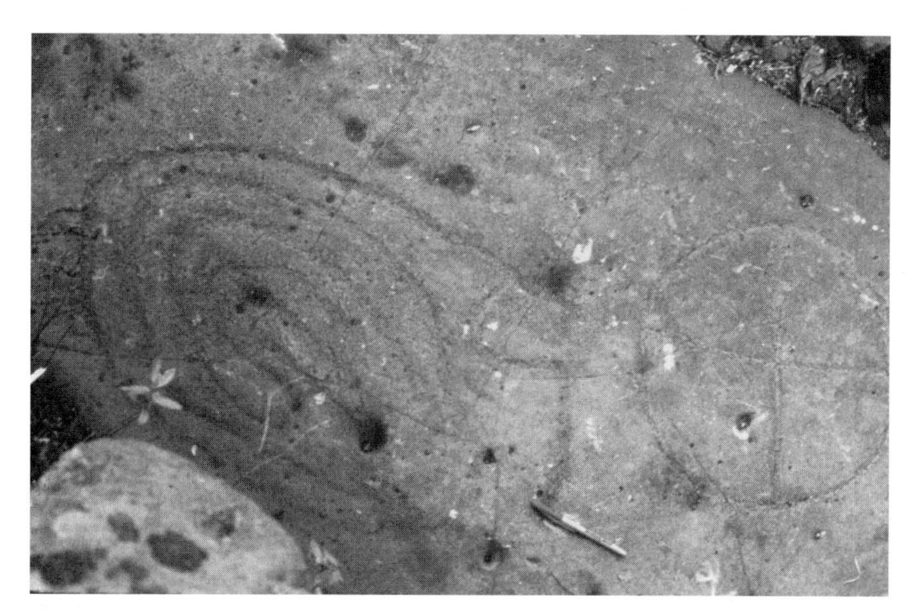

32. The ⌖ on Cerro Gordo (right)*, located high on a mountain north of Teotihuacán, was gouged over in modern times. (Photo by author)*

Architectural markers have precedent in other ancient cities. They survive in the Roman Forum, in Pompeii, and elsewhere as testimony to the process of city planning. Usually located at crossing streets, they consist of small, low, worked stones, often with Xs carved on them. Roman surveyors called them *cippi* (Latin for "posts") and writers tell us that they were used as landmarks and for indicating precise distances between places.

As in Teotihuácan, if you're going to build a great city—the place where time began—the need to make it resonate with the harmony of the cosmos might be expected to be a part of the plan. But why skew the plan 15.5° from the cardinal directions? Was time's beginning somehow reflected in that particular orientation? And, if so, how?

You can't imagine the unswerving passion of an astronomer confronted with the possibility of solving a puzzling problem like that. Decoding this mystery was right up my alley, an easy problem to solve—or at least a lot easier than my usual line of work, I thought. Normally I spent my time computing temperatures and pressures in the cores of young supergiant stars and measuring the rate of ionized gas flow from their surfaces.

Energized by my youthful enthusiasm, I went to the planetarium. I set the clock for 150 BC (the best date we could get from the archaeologists for the artifacts), marked out an artificial horizon on the dome wall, entered the alignment between the two ⊕ symbols acquired from our field measurements, turned on the projector, and then sat back to watch what happened. The sun, the moon, and the planets all moved predictably in their courses. The sun hit the alignment twice a year, but the equivalent dates didn't correspond to the solstices of Stonehenge fame. After the sun went down, it didn't take long to find a prime candidate—the Pleiades, or the "Seven Sisters." This conspicuous little star cluster was well-known to many cultures across time—from the daughters of Atlas in Western lore and *subaru* to the Japanese, to *tzab* (the tail of the rattlesnake) to the Aztecs. Dow had already suggested the Pleiades as a target for the ⊕ alignment. Located in our constellation of Taurus, this star cluster set along the skewed Teotihuácan east-west axis just about the time the main part of the great city was built. (It no longer hits that target because the precession of the equinoxes, or slow wobbling motion of the earth's axis against the background of stars, has shifted its setting point.) I made a few calculations and discovered that in the time and place of Teotihuácan, the Pleiades also passed the zenith, or the point directly overhead.

The Aztecs, who settled in the Valley of Mexico nearly a thousand years after the fall of Teotihuácan, named the Pleiades as one of their major objects of sky worship and they pictured the star group in their manuscripts. One chronicler tells us that they commenced the start of their 52-year religious cycle with a "binding of the years" ritual, in which priests ascended to the top of a special mountain—called the Hill of the Star and located on the outskirts of their capital city of Tenochtítlan—to sight the Pleiades as they passed the zenith at midnight. As the Pleiades passed overhead, he writes, they took it "as a sign to the anxious waiting multitude that the world would not be destroyed and that a new era would be granted to mankind."[5] Then came the clincher: I noticed that after the Pleiades disappeared from the sky in the spring, blocked out of view by the light of the sun for several weeks, they reappeared on exactly the same day that the sun passed the overhead point—an event identified with New Year's Day in Mesoamerica.

Those of us who live outside the latitude of the tropics have no sense of the phenomenon of solar zenith passage. In Chapter 6 we discussed it in con-

33a–b. Pecked cross-circle petroglyphs from Xihuingo, a colony of Teotihuácan. (Photos by author)

nection with the Javanese gnomon. Recall that as the sun travels on its daily journey, it gradually moves northward from winter solstice and reaches ever higher positions in the sky at noon, until it stands directly overhead. On that day the sun casts no noontime shadows. It equally illuminates all sides of a vertical pillar. Then it moves north of the zenith at noon and all objects again cast shadows, but this time to the south instead of the north. On June 21, the summer solstice, the sun reaches farthest north; then it turns around and begins to move back toward the south, passing the zenith at noon a second time. The dates of the two zenith passages and the interval between them depend on your latitude. Close to the Tropic of Cancer (latitude 23.5° north) the two solar zenith passages crowd in on the June 21 date from both sides. The farther south you go the further in time the zenith passages move away from the summer solstice. In Teotihuácan the two dates are, and have been for all of history (for they are little affected by precession of the equinoxes), sixty-eight days apart; they fall on May 19 and July 26.

As far as I was concerned I had solved the mystery of the ⊕: these markers were used to lay out the city in accordance with a principle ordained by the gods in heaven. They fixed the cosmic axis and they timed it with the movement of the sun (at the zenith) and the stars (the Pleiades).

Other decoders of the mystery of the Teotihuácan pecked cross-circles had proposed explanations far wackier than mine. One imaginative engineer, for example, argued that the ⊕ were geodetic markers "to locate man in space and time—to fit him into the cosmos."[6] What made his theory so eye-catching was that he even managed to find the dimensions of the Egyptian pyramids, the circumference of the earth, the orbits of the planets, and several universal mathematical and physical constants—including pi, the golden ratio, the natural base for Napierian logarithms, and the speed of light—all hidden away in the Teotihuácan grid system. As one reviewer commented, "[T]he way to success is to tempt the public with secret information."[7]

Closer to sanity, geographer Vincent Malmström argued that the alignment between the two Teotihuácan ⊕ symbols was a sun alignment, because one of the sunset dates that fit (August 12) is the day when the sun passes the zenith, not of Teotihuácan but of Izapa, a Maya site on the Pacific coast of Guatemala. There, he argued, the Mesoamerican calendar originated. When this calendar was imported to Teotihuácan, its place of origin was commemorated by the city's alignment. Aside from the fact that little evidence attributes the origin of the calendar to that particular place, I have always found his argument a bit contrived. But I suppose the Teotihuácan people, like the Aztecs of Tenochtítlan, were entitled to hold some ancestor responsible for time's creation.

There's a fable by Aesop about a man who marks the way to a treasure he has buried at a secret location by using his hatchet to put tiny nicks on tree trunks to preserve the trail. But he is foiled by a rival who does him in by putting similar nicks on practically every tree in the forest. Loss of treasure is not unlike loss of hypothesis. (If you guessed that more tick marks equal more ⊕ symbols, then you have captured my analogy.) What buried my orientation hypothesis (and most other orientation theories) was the rapid-fire discovery of hosts of pecked cross-circles both in and around Teotihuácan. A marvelous one was found in the Ciudadela (Figure 30), a large complex of buildings located half a mile south of the Pyramid of the Sun; three more were excavated out of the floors of other buildings that flanked the Street of the Dead. Three were found by hikers on Cerro Chiconautla (Figure 32), a flat-topped hill ten miles southwest of the site; another turned up on top of Cerro Gordo eight miles due north. By 1980, two dozen ⊕ symbols had been tallied; and by 2000, the count had risen to nearly fifty. Swiss archaeologist Matthew Wallrath charted another forty of them at Xihuingo, a Teotihuácan colony located twenty-five miles northeast of the great city along the obsidian trade route (Figures 33a and 33b are examples). He thought Xihuingo was a place deliberately built by astronomers to study and explore alignments, a kind of "Teotihuácan ⊕ Institute of Advanced Study"—a bit of a stretch for me. Then in 1984, while clearing the south platform of the Sun Pyramid, archaeologist Eduardo Matos Moctezuma—the excavator of the Templo Mayor whom we met in Chapter 6—unearthed several more, along with a host of other quadripartite designs.

The importance of the ⊕ symbol in and around Teotihuácan cannot be dismissed. Whatever it meant, the paramount foundation culture in all of Mesoamerica invested a huge quantity of energy to reproduce it. Just from our sampling, we calculated a density of 800 of them per square mile in Xihuingo alone, which is not exactly downtown Teotihuácan. Then there are the petroglyphs crowded along the south platform of the Pyramid of the Sun. Matos and his co-workers, headed by archaeologist Ruben Cabrera Castro, dug up forty-four symbols (not all of them qualify as a ⌖, as we'll see later) spread over a stuccoed floor space of 500 square yards. Based on my own experiments with the time it takes to hammer out the design with a percussive stone instrument (no, we didn't try that on the floors of buildings at the archaeological ruins!) and assuming that what Matos found is representative of what may once have existed throughout the city, I calculated that the people of Teotihuácan would have invested what amounts to 200,000 person-hours of labor executing them—a lot more time than even the most meticulous architects might have cared to consume in making benchmarks.

Cognitive dissonance. That's what psychologists call a mental conflict that happens when new information comes along that challenges your beliefs or assumptions. Symptoms include a feeling of unease or tension. For temporary relief, try a defensive maneuver: ignore the new information or try to explain it away—anything to preserve order in your ingrained conception of the world. Afflicted with the barrage of discoveries of new ⌖ symbols, I definitely was suffering from cognitive dissonance. Initially I wasn't even willing to think that I might be wrong.

Eventually I had to face it: no orientation hypothesis could withstand the assault of the ⌖. The proliferation of these discoveries didn't mean that some of the ⌖ weren't architects' benchmarks strategically located for the practical and religious function of astronomically laying out the Teotihuácan grid (as in the analogy to nicks on trees in the fable), but it certainly did imply that if I was looking for an overarching theory—a theory to explain *all* ⌖—I would need to look elsewhere.

While browsing the library of Mexico City's National Museum of Anthropology, I came across Mexican poet Vicente Riva Palacios's magnificent multivolume 1889 work *Mexico a Través de los Siglos* (*Mexico Through the Ages*). It was in Alfredo Chavero's chapter on ancient Mexican calendars where I glimpsed the words that make up my second epigraph to this chapter. They offer a tantalizing description of an elusive artifact found "on the northern frontier of our country" that sounded exactly like the ones I'd been researching in Teotihuácan. I followed Chavero across the shelves to an earlier 1886 publication, which contained a drawing of the artifact that left no doubt about my suspicion. About that time, one of my Teotihuácan consultants thought

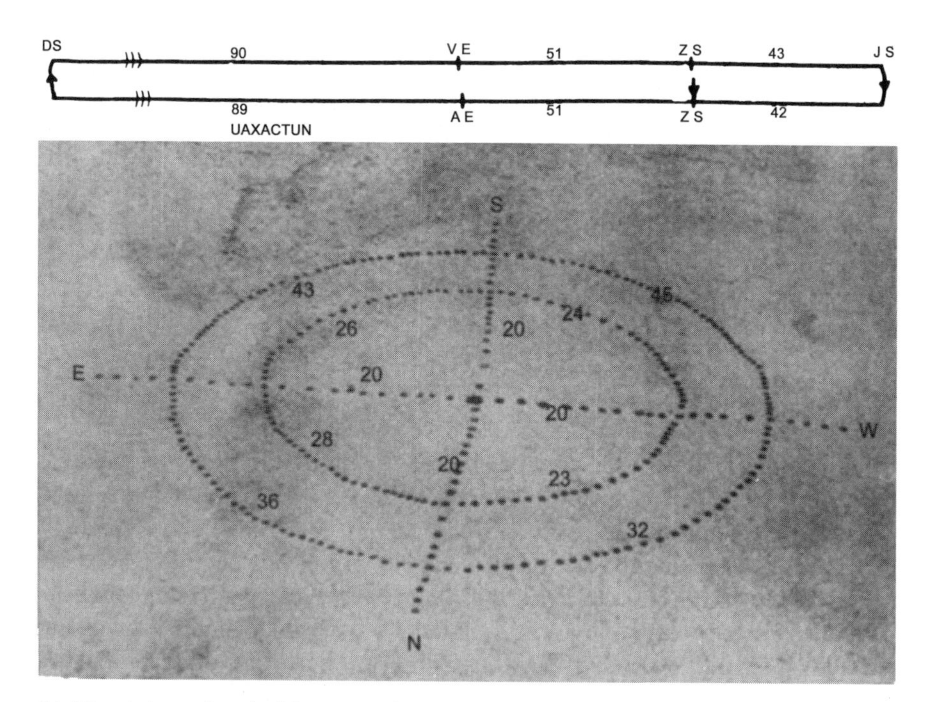

34. The ⌖ located in the Maya city of Uaxactún. (Photo by author)

he remembered a pecked cross-circle discovered somewhere in the Maya region (Dow had even mentioned it). When I searched the old Carnegie Institution's archaeological reports, I found the description given in my third epigraph. It tells of a ⌖ at the ruins of Uaxactún, one of the oldest Maya sites in the Petén rain forest of Guatemala, located more than a thousand miles east of Teotihuácan. An accompanying photo (Figure 34) taken in the 1940s by Harvard archaeologist Ledyard Smith revealed an almost identical design that had been pecked into the fourth-century AD stucco floor of a building Smith had excavated.

I had never paid much attention to the peck marks themselves—their arrangement, their spacing, their number—even though both Chavero and Smith had remarked about these matters in their written descriptions. Was I too blinded by the quest for alignments? by questions about directions of the lines connecting various markers? These were *my* questions—the astronomer's questions. As I look back I think I already knew that the first Teotihuácan petroglyph had a 10-4-4 axial pattern (10-5-5 if you count the holes that mark the intersection of circles and cross elements); and that this pattern followed exactly Smith's description of the ⌖ at Uaxactún. In hindsight I suppose that, like Chavero, I must have been mindful that keeping time could have had something to do with the mystery of the ⌖. But no excuses and, notwithstanding my love of numbers, for some reason I had just plain avoided them when it came to the ⌖. When I looked more closely at the count on the Chavero design, I began to take more seriously the idea that they might have had something to do with counting time. Using a magnifying glass, I counted 20 holes per quadrant on the inner circle, for a total of 80; 25 per quadrant on the outer circle, totaling 100; and based on what I could see (a portion of the stone had been broken off), 20 on each axis, totaling 80. The grand total was 260.

Regardless of where and when it happens, discovery is a process that begins when the human imagination recognizes a likeness or similarity in what once were two disconnected phenomena. Newton saw a falling apple and connected it to the motion of the moon, and Einstein tied movement through space to a perceived flow of time. As I mentioned in Chapter 7, I think one of the greatest discoveries of ancient Mesoamerican timekeepers lay in the recognition of one of nature's magic numbers—260—the one that brought to light once unrelated aspects of a functioning real world, such as the cycle of the planet Venus and pregnancy, the moon and the body count. Finding 260 is a chapter in the story of the search for intelligible order in a seemingly chaotic universe that we all share as rational, sentient beings.

Numbers—documentable, repeatable, calendrical numbers—are good reasons to probe the calendar hypothesis: maybe ⌖ symbols are devices that depict ways of counting units of time. But what do we know about what the sages of ancient Mexico thought of the passage of time? Clearly, I needed to pay more attention to documents that dealt with Mesoamerican timekeeping.

The Fejérváry Mayer map (discussed in Chapter 7 and shown in Figure 27) had taught me a lot. It showed me how ancient Mesoamericans celebrated the repetitive flow of time and the completion of its many cycles: the day is the passage of the sun along the main axis of the cross; the 260-day biorhythmic cycle resides in the sequence of dots that make up the border; the passage of the year is encoded in each intercardinal leap; and a bundle of 52 years—the time it takes the 260- and 365-day calendars to realign—comes with thirteen traverses via the intercardinals. This "binding of the years" cycle may have held special importance at Teotihuácan, for there both the Pleiades and the sun crossed the zenith, and the Pleiades, according to the Aztecs, signaled the day to celebrate the 52-year ritual.

There are other designs just like Fejérváry One. An almost exact replica appears in the Madrid Codex, a fifteenth-century Maya document. The Maltese-cross design also appears in ancient Mexican manuscripts, where it seems to signify the completion of time's cycle. I recall it most vividly in the Codex Borbonicus (dated to around the time of first Spanish contact), where it appears on the doorjamb of a temple where the New Fire ritual that completes the 52-year cycle is being carried out. The gods of the four directions, their eyes marked by that identical symbol, meet at the center. Each carries a bundle to symbolize the years; they are just about to thrust them into the fire in the temple's altar. The altar is decorated with the same circle/step motif we find on buildings at Teotihuácan. One Maya hieroglyph that also has the shape of a Maltese cross—the katun, or 20-year cycle—is translated as "it is completed." The floral symbol also represents the last of the twenty named days in the Aztec time count.

Fejérváry One reminded me of a design I had seen pecked into the floor of a building just across Teotihuácan's Street of the Dead from the first ⌖. This petroglyph has a shape very similar to Fejérváry One, except that there are three pecked Maltese crosses, one inside the other. I counted the peck marks on the outer perimeter of this triple-cross petroglyph: 260. This petroglyph, along with several others like it, added more weight to the hypothesis that the designs had some connection with time. But one question inevitably leads to another. Were the ⌖ symbols timekeeping devices—mechanisms like clocks or calendars specifically designed to measure and mark time? Or were they records, or perhaps simply reminders of the ritual importance of the completion of time's cycle?

I have always had a fascination with clocks. I remember as a child (I could not have been more than four years old) cutting out cardboard circles and making clock faces out of them. I would fasten cutout arrow-shaped hands to the clock face with a bent paper clip. Sometimes I left out a number, wrote a number backward, or laid out the number sequence in counterclockwise instead of clockwise order. My grandmother would correct me. I didn't know how to tell time. In fact, I learned to do so in stepwise fashion by perfecting my cardboard clock faces. I also recall being preoccupied with getting the correct number of stripes (13) and stars (then 48) on the U.S. flags I drew. The rectangular 6-by-8 stellar pattern was a lot easier to configure than the contemporary alternating pattern of sixes and sevens. Some of my earliest masterpieces show I had some trouble getting it right.

I resurrect these childhood memories because I noticed that the tallies on many ⌖ symbols deviated from the 20/260 arrangement. What if some of the crosses were training devices made by novices, or perhaps intended for instruction in the making of calendars? The large number of ⌖ symbols found crowded together on the floors of buildings lining Teotihuácan's Street of the Dead look like so many scratchings on an unerased classroom blackboard at the end of a busy day; they could have served just this purpose. Some graffiti etched on the walls of buildings at the Maya ruins of Tikal (twelve miles due south of Uaxactún) years after its abandonment appear to be quick sketches with imperfect attention to detail—like my early clocks and flags. In fact, they look as if they might have been made by kids (most of them are just a few feet above floor level). And although they attest to the universal importance of the ⌖ symbol, they offer us about as much precise information on the pecked cross-circle petroglyph as a *New Yorker Magazine* cartoon picturing a microscope can enlighten us regarding the principles of optics.

How else to explain the wide variation in counts on the ⌖ symbols? Farming is one good reason for standardizing a calendar. Agriculture still lurks behind Arizona's and (portions of) Indiana's reasons for opposing Daylight Saving Time. Farmers don't like the idea of "milking cows in the middle of the night," as one of them put it when he experienced pitch-darkness at 7:00 AM on autumn mornings under the rule of DST. Last frost comes in early May in the area where I live—that's when local farmers put in their peas. First frost happens at the beginning of October. From then on its green tomatoes, provided you get them in soon enough. Travel 500 miles south of here and the planting-to-harvesting interval is nearly doubled. Likewise, rains begin in earnest in Teotihuácan in the month of May, just about when the sun passes the zenith. They max out in July and diminish in late September or early October. A few hundred miles north, around the Tropic of Cancer, the rains begin as early as late April, reach a peak in August and September, and don't diminish until November.

Travelers, traders, exporters, importers, and most definitely farmers—there were plenty of each of them around in the heyday of Teotihuácan—would have been well aware of the warp in time among the cities and towns that made up their extended world. Pre-Columbian documents are filled with farmers' almanacs showing pictures of offerings being made to the maize god and the rain god, who is often shown shooting his bolts of lightning and planting, scattering, or sowing seeds with his planting stick. If ⊕ were about counting real time—especially the time of year to perform the rituals that must have attended the anticipation of rain and the sowing of crops—then, with the possible exception of the paramount 260, we would not expect the numbers that make them up to be the same. Getting the right time was certainly beginning to look like the answer to the mystery of the ⊕.

J. Charles Kelley was the archetype of the old-school archaeologists who had worked the Mexican border all the way back to frontier days. Imagine the goateed colonel on the Kentucky Fried Chicken logo, give him a heavy Texas twang, put an Indiana Jones hat on his head and he's ready for the trenches. When I first met him, I found the white-haired septuagenarian fit, even though he walked with a limp he had acquired from years of jumping into numerous hip-high test pits he'd dug all over the desolate *pedregal* (literally, "rock pile") of northwest Mexico, his area of specialty. His adoring wife and field companion Ellen called him "J. Charles," drawing out the last syllable with the same Texas twang. The moniker caught on with me and my students when we traveled with him to Alta Vista on the June solstice of 1977 to glimpse for ourselves his latest discovery. He had told me in an urgent phone call a month earlier, "This is gonna knock your socks off!" That discovery would turn out to be neither a cache of jade nor a skeleton, but an alignment.

Alta Vista is an ancient settlement (Kelley calls it a ceremonial center) located in a wide valley a few miles south of the Tropic of Cancer. That it offers no obvious defensive or ecological advantages (the nearest water supply lies a considerable distance to the east) is enough to make one wonder why anyone would care to live there. The place was rife with Teotihuácan ceramics and, by studying the sequence of styles he'd excavated, Kelley had pretty much documented that Alta Vista was a Teotihuácan outpost dated to AD 450–650. Teotihuácan lies an impressive 400 miles—that's about a month on foot over rugged terrain—to the southeast.

Our late June tour with J. Charles started at Alta Vista's largest building, the Hall of Columns, which he had named after the twenty-eight round pillars that once supported a roof. The corners of the Hall of Columns line up perfectly with the cardinal directions. Along the outside of this colonnaded building, Kelley had excavated an elaborate masonry-walled structure. He called it the Labyrinth because of its many bends and turns. A pathway about a yard wide led us out of the Labyrinth in an easterly direction. Standing at its exit point, we found ourselves facing due east squarely in the direction of the most prominent landmark—a sharp pointed peak named Picacho el Pelón situated seven miles out on the horizon. Locals had told the archaeologist that their ancestors worshipped Picacho because it was the source of their precious blue water (a natural spring emanates from its base) and their precious blue stone—the turquoise that comes from Picacho's Esmeralda Mine. The adjacent village is named Chalchihuites, or "blue stone" in the local dialect.

J. Charles whipped out a photo he'd taken on the previous March 20 equinox from the spot where we were standing. It showed the sun rising precisely over the pinnacle of Picacho. Impressive, but my socks were still on—until we climbed Cerro El Chapin, a 300-foot-high, quarter-mile-long plateau located nine miles southwest of the ruins. Once on its summit, J. Charles and a local guide used machetes to cut a path through prickly pear, mesquite cactus, and thorny brush to a flat rock clearing on Chapin's eastward-looking eminence. There, as he displayed a wide, self-assured Kentucky colonel grin, I beheld the most exquisite looking Teotihuácan-type ⊕ that I had ever seen (Figure 35). I immediately dropped to my knees and started counting: inner circle, 80; outer circle, 100; axes, 80; total, 260. Fifty yards to the north lay a second, almost identical design. Since Durango, Mexico—about where we were located on the map—fit Chavero's vague description of the place "on the northern frontier," where the earliest reported ⊕ was located, I took out my notebook and compared his 1886 sketch with the object at hand. Even though the tallies were identical, the two were clearly not the same object.

When he surveyed the plateau for ceramics, J. Charles noticed that the axes of both petroglyphs seemed to direct the viewer's eye toward Picacho.

35. Archaeologist J. Charles Kelley points out an alignment as he stands on a ⊕ at the Tropic of Cancer. (Photo by author)

Once we bushwhacked the cactus in the foreground we quickly verified his observation. But the wise old archaeologist had brought us here on the eve of the June solstice for another reason. To judge by the Mexican government topographic maps, the line extending from the top of El Chapin to Picacho lay very close to the point where the sun rises on the first day of summer. We would anxiously await the next morning to see for ourselves whether it was true.

Camping out under the stars with J. Charles was an unforgettable experience. He regaled us with campfire stores about archaeology's early days—unearthing treasures of turquoise, shooting rattlers, being held up by banditos. We listened enthralled, downing the cold, bottled brew we had carted in our packs to the top of the hill, nonfiction and fiction blending as the starlit night advanced. This evening's repast was concocted—using J. Charles's recipes based on mathematical permutations of numbered units of diverse tinned foods—by combining two cans of Dinty Moore beef stew, one can of Armour corned beef hash, and three cans of Brand X baked beans, finished (read "glued together") with a tiny tin of Underwood deviled ham spread. The opaque, sediment-rich coffee we washed it down with was a far cry from today's frothy concoctions from Starbucks. But the company more than made up for the meal.

Next morning we awoke before predawn twilight, excited to set up our surveyor's transit and position it on the crossing point of the axes of each design. Our Nikkormat loaded with 35 mm slide film for documentation, we eagerly awaited the first glimpse of the great luminary. Not a cloud in the sky, the silent air glowed with increasing brightness, silhouetting the purple line of hills to the east. In the midst of the Sierra Madres, needle-like Picacho protruded. Then first light—a dazzling yellow ray glinted off the very tip of Picacho Peak and flooded the valley heralding the season's longest day—the day of zenith passage at the Tropic of Cancer. What happened to my socks?

Let's put the pieces of the puzzle together: Alta Vista was an outpost of Teotihuácan, the most influential city in early Mesoamerica, and positioned 400 miles northwest of the great city at a nondescript remote location close to the Tropic of Cancer—but *carefully* positioned, with its main temple cardinally aligned. A ceremonial pathway leads outward from the temple in the direction of a very important peak—a turquoise/water mountain that happens to lie due east, the place where the sun rises at its seasonal midpoint, the equinoxes. A pair of ⊕ symbols located on a hill overlooking the site are aligned to the same eastern eminence, except that, seen from their location, the sun rises over that same peak on the June solstice, the maximum northerly extent of horizon that marks the sun's annual journey, and—here's the bonus—the same day that the sun reaches the zenith at noon, the one day of the year when it casts no shadows. This event happens in no other latitude in the northern hemisphere. Finally, the arrangement of the number of elements that make up the design also seems to have some significance. They are set out in multiples of 5, counts of 20, and they sum to 260, the most important cycle in the Mesoamerican calendar.

Again I thought about Fejérváry One—that microcosm of ancient Mesoamerican space-time—the ancient Mesoamerican "map" that offers a place for everything and puts everything in its place. The Alta Vista complex seemed a macrocosm of the same picture. The temple lay at the center of coordinates marking out the four directions. Around it moved the day—the sun—time's direction giver, periodically passing the mountain of sustenance. I also thought of the skill and patience that must have been required to fix the Alta Vista macrocosm in place. Two horizons with different elevations viewed from two observation points miles apart—one aligning with the solstice from Chapin, the other with the equinox from Alta Vista—and both over the same mountain. Yet the double alignment seems to have been engineered to work perfectly. How did they manage it? My guess is that the Chapin site was selected first, then the astronomical and topographical considerations determined where the ceremonial center would be located. But first the Teotihuácan astronomers needed to find the Tropic of Cancer.

Tropic comes from the Greek word *trope,* or "turn." When Ptolemy of Alexandria identified the parallels on the earth we still call the Tropics, he did it by studying the lengths of noontime shadows on sundials.[8] He defined the region between the two Tropics as *amphiskian* (*skia* = shadow, *amphi* = on both sides), because the shadow of a vertical post is cast to the north for part of the year and to south for the remainder in this region. He named the Temperate Zones *heteroskian* (*hetero* = on one side). Knowing how dials behaved in different places at known distances apart, Ptolemy was able to determine approximately where the dividing lines for the climate zones lay. This calculation would entail a long-range program of repeated observations around the time of the summer solstice at many test sites near the Tropic. He located the Tropic of Cancer in "Cinnamon Country," far up the Nile in today's Sudan.

Interestingly enough, we located another pecked cross-circle petroglyph thirty-five miles to the northwest of Alta Vista, or about twenty-five miles north of the Tropic, that mimicked the Chapin pair. Could this ⊕ have been a relic of an earlier attempt to find the Tropic? With a post set straight up in the ground or by passing the light of the noon day sun through a carefully aligned vertical tube, an observer could pin down the location of the Tropic to within fifteen miles or so. I calculated that the Alta Vista pecked cross-circles were twelve miles south of the location of the Tropic in AD 650. Next, one would need a noteworthy permanent landmark to register the solstitial sunrise as well as an appropriate backsight from which to make the annual observations and conduct any ritual that might attend them. Any alignment parallel to the line from Chapin to Picacho would work, but once I saw the phenomenon I was convinced that few would have made the event seem as powerful as the one that confronts the viewer perched atop Chapin on summer solstice morning.

The next step would be to lay out the equinox line in order to determine where to build the Sun Temple. But determining the equinox is not so simple. If the first days of autumn and spring fell exactly midway in time between the beginnings of summer and winter, an observer could simply count days from the solstice and put markers in place at the midpoints in time between the first days of summer and winter. But they don't. Our seasons are not equal; the winter-to-summer interval is about six days longer. To minimize the error, you could make an averaging process consisting of double observations from opposite solstices.[9]

Whether these techniques or some other procedure was employed, the whole process of laying out the Alta Vista alignments at the Tropic of Cancer must have required great organizational capacity and a good deal of patient observing, although a technology of only modest sophistication would have been necessary. Architects capable of planning and constructing Teotihuácan would surely have been equal to the task of developing and producing the astronomical alignments we measured at this unique site at the Tropic of Cancer.

The most reasonable conclusion we can reach about the ⊕ at the Tropic is that these artifacts have something to do with space—the setting out of important directions—*and* time—the tallying of the temporal rounds that make up the calendar. The whole scheme seems to be an attempt to map out the space and time that envelop Alta Vista, which fits into the system via its special orientations. The mountain with its precious substances is likewise given its appropriate place in the scheme. The horizon, just as in Fejérváry One, marks the midpoint of time's seasonal cycle along with one of its endpoints. Conceivably the other endpoint—the December solstice—may have been marked as well, although we would need to scour dozens of square miles of *pedregal* to find the ⊕. The sacred mountain that issues its vital substances is doubly marked, firmly attached to the path of time. And finally, the count of time is registered in microcosm around the periphery of the ⊕ as well as along its axes.

The archaeological record tells us that the whole scheme was dreamed up by people from Teotihuácan. They were a sophisticated people with an influential style worth imitating, who had built a cosmically aligned city. Their influence in art and architecture had spread across Mesoamerica (Teotihuácan artifacts have been found as far south as El Salvador and as far north as Arizona). Like the Greeks, they sought out the place where the sun turned around. Whether they believed doing so would give them a measure of control over the gods who moved time or would simply serve as a means of gaining greater insight into anticipating how their gods would behave, who can say? It certainly gave them control over time and space. Given our knowledge and admiration of the skills and interests of the architects of Teotihuácan, we might expect people in other parts of Mesoamerica to set up solar alignments and time counting—even the Maya far to the east.

It took me a long time after I first came across Ledyard Smith's description of the ⊕ (the third epigraph for this chapter) to get to Uaxactún in the Petén rain forest of northern Guatemala. Since the 1980s the jeep trail has been improved so that today even casual tourists who choose to spend an extra day at the ruins of Tikal can journey, at least during the dry season, twelve miles north to its ancient predecessor. Inscriptions on a Tikal stela tell us that the ruler of Tikal, Toh-Chak-Ich'ak (True Great Jaguar Paw), wrested power from

Uaxactún in a great battle in AD 378. When I finally reached Uaxactún and got my first look at the ⌖ Smith had excavated some thirty years before, I was surprised to see what a pristine specimen it was (Figure 30f). Carefully hammered out on a section of the 10,000 square-foot floor of structure A-V, it was accurately datable to no later than AD 378 by the inscription on the stela that rests on the same floor.

What's a Teotihuácan ⌖ doing in the middle of the land of the Maya a thousand miles away? Foreigners were in their midst, the Maya say. "[T]he arrival of strangers . . ." read the hieroglyphs on a Maya stela at the ruins of Uaxactún dated to the late fourth century. Epigrapher David Stuart translates the rest of the passage: Siyah K'ak ("Fire Is Born") was the name of their leader and he came . . . "from the west" . . . "to oversee it" as it was installed.[10] Whether Siyah K'ak was an invader or came in peace is not clear, but the influence of Teotihuácan flourished all over the Maya area: ceramics, Teotihuácan-style paintings, decorative motifs, emblems, architectural styles—and the way they reckoned time.

The count: double circle, twenty marks per axis. But there the resemblance to Teotihuácan ⌖ symbols ceased. Even from a casual glance it was easy to see that although they were evenly spaced, the peck marks on the southern side of the ⌖ were neatly cramped together—as if the maker had been required to squeeze in the appropriate number of them. I counted: 88 on the southern half, but only 68 on the northern half of the outer circle. If you'll recall, 68 was the number of days between the two annual passages of the sun across the zenith at Teotihuácan. The corresponding interval is 85 at the more southerly latitude of Uaxactún. That's close to the tally on the southern half of the outer circle. The number of dots on each half of the inner circle, 51, replicates exactly the interval between equinox and zenith passage at Uaxactún. At Teotihuácan the corresponding interval (60 days) is marked out by the east-west axis of the Teotihuácan grid, the alignment we believed to have been fixed by our first two Teotihuácan pecked cross-circles. In other words, once the sun reaches the vernal equinox, or true east-west, it takes 2 x 20, or 40, days to align with the pecked circles on the Teotihuácan grid. From that point, add 20 days to reach first zenith passage at Uaxactún.

This curious mixture of precise information about Teotihuácan and Uaxactún seasonal intervals, both fixed by nature and each represented in a

Maya petroglyph, tells us that the two cultures had intimate contact with one another regarding a matter of great concern—the performance of sacred rites in accordance with the setting of nature's timepiece. I think that, like the people of Alta Vista on the northern Tropic, the Maya of Uaxactún were adopting "Teotihuácan Standard Time." Further proof came when we measured the axis of the Uaxactún petroglyph and found that it aligned with the 15.5° skewed grid system of Teotihuácan—the place where time began.

The sky and the calendar—space and time—may have been the underlying motives for making the ⊕. Taken together they offer some insight into how the designs were put there, from the *pedregal* to the Petén, and in and around the great ancient city that gave rise to their invention. But the data on alignments and tallies don't tell us who used them. Who sat around these ancient petroglyphs? Who stood on them? My own background and training wants to make them people like me, stargazers blissfully concerned with marking out the universe of space-time. But the ⊕ symbols also could have been oracles visited by shamans trying to read the omens issued by the dawning of the sacred days of the 260-day count.

There is in Xihuingo a magnificent ⊕. It is carefully positioned on a cliff at the extreme southwest end of a flat-topped hill that overlooks the ruins. Its axis is oriented precisely toward Cerro Gordo, the large mountain overlooking Teotihuácan on the north. The view along the axis of the ⊕ passes through a saddle-shaped feature framed by a pair of low hills in the foreground. Cerro Gordo can be seen majestically protruding from the center of the saddle. The visual effect gives the impression that careful topographic considerations dictated the selection of the rock where the people of Teotihuácan would peck their symbol—more a geographic than an astronomical pointer. But I prefer to think of this ⊕ as a device intended to fix one's gaze upon the place where time was born—perhaps the gaze of one destined to engage in a ritual at precisely the appropriate time. Could they have been healers? We know of at least one contemporary curing rite in highland Guatemala in which balls of copal incense are distributed in a pattern that resembles a pecked cross-circle design and is thought to possess curative powers. Maybe that's why so many ⊕ symbols appear in porticos and apartment complexes in the personal space of Teotihuácan householders.

Or are we just playing games? There is a game called *patolli* that is still played by the Indians of northwest Mexico. They play it on a "board"—actually a pecked, scratched out, or painted surface, usually a flat rock or a stucco floor. We find *patolli* boards all over Teotihuácan, many of them on the same floors that contain ⊕ symbols. More than a dozen of them intermingle with the ⊕ on the south platform of the Pyramid of the Sun. There are several at the Maya ruins too. The playing surface usually takes the shape of a square, often with flying corners—like a swastika. As in the east Indian game of *parcheesi* (*pachisi*), players move markers along the track; the number of spaces they move is determined by casting lots. Likewise you can capture or kill (they say "eat") one of your opponent's pieces by landing on the same space it occupies. Indeed, some anthropologists think *parcheesi* and *patolli* are so alike that they surely must have had a common origin. Here is another similarity reminiscent of the arrangement of elements that we consistently find on the ⊕: everything is arranged in fives. One sixteenth-century Aztec chronicler who saw the game played during early conquest times says that the dice (actually beans with small white dots painted on them) were numbered one through five, and ten; then he adds the rather odd statement that "when the painted number was five, it meant 10, and ten meant 20."[11]

We think of games as pastimes; but my point of view changed when I came across a detailed discussion of the circumstances surrounding the way they played the game of *patolli* in ancient Mesoamerica. The chronicler who witnessed a group of men seated in one of the rooms of a ruined Aztec temple deeply engaged in *patolli* has left us a detailed description of how it was played. As you read the lengthy passage I am about to quote, please keep in mind that it was penned by the hand of a Spanish priest bent on converting what he believed to be a pack of heathen savages, so he isn't about to cut them much slack.

> The gamblers dedicated to this game always went about with the mats under their armpits and with the dice tied up in small cloths, like some gamblers today, who go from board to board carrying their cards inside their hose. They were revered as gods, as it was believed that they were mighty; and thus when they played, [the people] spoke to them as if they possessed reason

> or intelligence regarding the request made to them. That they spoke to them and begged them to be favorable to come to their aid in that game, does not surprise or astonish me. Though these were people who were less alert than our own people [I must admit that] there are Spanish Christians who (though pretending refinement) when the cards are being dealt demand of the card a good number and good fortune and if this is not obtained after having "worshipped the cards" (if thus it may be termed), with the cards in their hands they give voice to a thousand blasphemies against God and his saints. In this way the natives spoke to the beans and to the mat, uttering a thousand loving words, a thousand compliments, a thousand superstitions. After having spoken to them, they placed the painted mat and the small case containing the implements of the game in a place of worship. They brought fire, cast incense into the flames, and offered their sacrifices in the presence of the implements, placing food before them. When the ceremonial gift had been delivered, they went off to play in the most carefree manner.[12]

Only at the end of this passage does it become clear that these "heathens," as the chronicler calls them, were not just idling away the time: "speaking to the beans" to answer "a request made to them" sounds more like the language of a shaman than that of a riverboat gambler.

I am reminded of an anthropologist's description of the duties of the contemporary Maya daykeeper. He meets with his client over a table—usually a flat rock located at the edge of the village. He removes all the divining apparatus—candles, incense, precious stones, crystals and beans—from his sacred bundle and lays them out on the rock, every piece in its proper place and all arranged according to the cardinal directions. After setting his crystals and beans in different number combinations on the table the client asks a question, will my daughter's unborn child safely enter this world? The shaman addresses the little heaps in turn: "Come here 1 Quej [one of the day names in the 260-day calendar still in use], you are being spoken to."[13] Then he begins counting the days according to the groups of seeds and crystals. He reads them; he talks to them, one after another. He moves them in a counterclockwise fashion until he feels "blood movement"—confirming that the answer will come through the "feeling in his blood." Whatever we might say about the school of specialists who poured over the forty-four symbols engraved on the stucco floor on the south side of the Teotihuácan pyramid, or the power of a pair of pristine ⊕ symbols far to the north in Alta Vista, or the cross-circle petroglyph in distant Maya Uaxactún, I think we must conclude that discourse with the gods had something to do with all of them.

NOTES

1. James Dow, "Astronomical Orientations at Teotihuacan: A Case Study," *American Antiquity* 32 (1967): 326–334 (after a verbal description by James Bennyhoff, 1963).

2. Alfredo Chavero quoted in Vicente Riva Palacios, *Mexico a Través de los Siglos* (Mexico City: Editorial Cumbre, 1889), 1:737. See also Annales, Museo Naciónal (Mexico) 3 (1886): p1.c2.

3. A. Ledyard Smith, Uaxactún, Guatemala. Excavations. Carnegie Institution of Washington Pub. 588 (1950), 21–22.

4. Rene Millon, *Urbanization at Teotihuacan,* 2 vols. (Austin: University of Texas Press, 1973), 1:42.

5. Bernardino de Sahagun, *Florentine Codex: General History of Things of New Spain* 3, tr. A.J.O. Anderson and C. E. Dibble, *Monographs of SAR* (Santa Fe: School of American Research, and Ogden: University of Utah Press, 1978 [1585]).

6. Peter Tompkins, *Secrets of the Mexican Pyramids* (New York: Harper, 1976). I refer specifically to the analysis by Hugh Harleston on page 241.

7. Raymond Sokolov, *New York Times Review of Books* (November 21, 1976), 8.

8. *Ptolemy's Almagest,* tr. G. J. Toomer (New York: Springer, 1984), 2:6; see also James Evans, *The History and Practice of Ancient Astronomy* (New York: Oxford University Press, 1998), 62.

9. Furthermore, the annual task of marking the date of arrival of the sun at the solstice by watching sunsets is also quite formidable because the sun slows drastically as it approaches and recedes from its horizon extremes. The work of ethnologist Ruth Bunzel from the 1930s provides us with concrete evidence on how contemporary native people who live to the north of this region accomplished the task (cf. Ruth Bunzel, "Introduction to Zuni Ceremonialism," *Bureau of American Ethnology 47th Annual Report* [1932], 467–544). She refers specifically to the *pekwin* or shaman who, around the time of the solstice, retreated to his sanctuary high atop the mesa to observe the position of sunrise and sunset over distant landmarks. Ceremonies were held for several days around each solstice and the announcement of the actual arrival of the sun at its extreme positions was made eight days before. They planted prayer sticks precisely on the solstitial date. On another occasion she tells us that the warning interval was ten days. The shaman must have struggled continually with the problem of predicting precisely when the sun would come to its annual northern standstill. Finally, refraction of the sun's disk by the earth's atmosphere plus a large share of cloudy days when you can't see the sun rise or set complicate matters

even more. Bisecting the tips of early morning and late afternoon shadows of equal length or sighting the Pole Star are alternative possibilities. There are references in the chronicles to ancient Mesoamericans having recognized the Pole Star. By making nightly observations, an observer would notice that all the stars circulate around the north celestial pole—the point in the sky marked closely by Polaris. With small corrections to allow for the movement of Polaris to either side of the pivotal point, they could mark the midpoint of the daily shift of the Pole Star and extend the line from the sky vertically downward to the horizon with a plumb line. This would give the astronomers a north-south line. The east-west axis could be obtained via a simple bisection of the north-south line using stakes and strings. Of course, all of this is speculation. Archaeologists haven't unearthed a shred of evidence that things were actually done this way. There are no techno-artifacts at Teotihuácan or Alta Vista. But the fact remains that they did it—and they did it with great precision.

10. David Stuart, "The Arrival of Strangers, Teotihuácan and Tollan in Classic Maya History," in *Mesoamerica's Classic Heritage: From Teotihuacan to the Aztecs,* ed. Davíd Carrasco, Lindsay Jones, and Scott Sessions (Boulder: University Press of Colorado, 2000), 465–513.

11. Diego Duran, *Book of the Gods and Rites and the Ancient Calendar,* ed. Fernando Horcasitas and Doris Heyden (Norman: University of Oklahoma Press, 1971 [1580]), 303.

12. Ibid., 304.

13. Barbara Tedlock, *Time and the Highland Maya* (Albuquerque: University of New Mexico Press, 1992 [1982]), 162.

TEN

LAST WORDS

Once upon a time a tribe lived deep inside a cave. Their existence was a strange one, for they were imprisoned, shackled by the necks and legs in a seated position for all of their lives, so that they permanently faced the illuminated wall at the back of the cave. A fire burning high on a wall behind them illuminated their view. On a rock ledge between them and the fire lay a road along which members of another tribe traveled, a road so situated that the shadows of the people of both tribes, the one moving and the other fixed, were cast on that same wall.

One day a member of the bound tribe managed to break loose his shackles. He rose up for the first time, turned around, and in an instant discovered the universe that lay behind him—a universe of moving three-dimensional images against an annoyingly bright background almost too painful to look at. But he persisted and, as he got used to the light and ventured closer to the moving images of the second tribe, vague outlines became discernible shapes—shapes that he realized resembled his own and that could be connected to the shadows he had watched on the wall for so long. He decided to follow the procession out of the cave, where he confronted another kind of light—a light even more awfully painful than the last. It was a blinding light, so penetrating that it drove him back into the semidarkness of the cave. Still he persisted and as he reemerged, this time more gradually, his eyes slowly become accustomed to the new light. It seemed to emanate from an extraordinarily bright object overhead. He discovered that this powerful light illuminated a vast environment of fields and mountains where other tribes lived and lakes and streams populated by other kinds of beings. Sometimes that light was periodically extinguished; then he beheld a black dome studded with countless bright points of light above his head. He learned to live in this world, all the while reflecting on how very different his new point of view seemed from that of the world he once perceived about him way down deep in his cave.

Eager to tell about his amazing ventures in this strange new world, one day he decided to return to his old habitat. Along the way he stumbled and fell, for he had difficulty readapting his eyes to what was now a dark, foreign environment. When he finally arrived, he excitedly revealed his new impressions of the phantom shadows his comrades had faced for all of their lives. He told them about the other people, places, and things he had witnessed and the bright light above that illuminated them.

Now write your own ending to Plato's celebrated cave allegory.[1] Are his cave-mates awed by the truth—a truth that lies outside their senses? If unshackled by their comrade, do they warily—or perhaps eagerly—follow him out of the cave into the new light? Or do they say his eyes were corrupted? Do they simply laugh at him? Once freed, might they seize and kill him for trying to corrupt them?

Different cultures ask different questions about the world around them. They seek different kinds of knowledge and, as we have seen, often they arrive at different kinds of truths. Here I'm thinking of culture as a body of beliefs people share; these beliefs can be social, religious, material, or any combination thereof. Culture can refer to the archetypal "other"—people who live or have lived in different places and times, such as the Maya, the Inca, the Dogon, or even Plato's imaginary cave people. Historical periods also manifest cultures that define and distinguish them from those of other times. We can speak of an Enlightenment culture or a medieval culture. And the specialties in which we are trained—what we call "disciplines"—are cultures of their own; thus, I can refer to the culture of a late nineteenth-century geologist as opposed to that of an eighteenth-century biological taxonomist.

Among the body of beliefs shared by each culture is a common sense, a take on what kinds of questions to raise in dealing with the natural world. An eighteenth-century taxonomist, for example, might ask, how is that plant used? but his modern descendant is more likely to query, what do its pistil and stamens look like? How the tribespeople in Plato's cave respond to the returning adventurer's description depends on how deeply the cultural tradition that gives rise to their common sense of the world around them is rooted.

In *Uncommon Sense* we have recounted a number of instances in which common senses collide. It happens when we persist in exlusively asking our own kinds of questions. Seeking an antidote to the contemporary habit of only searching for knowledge that leads to a single, fixed, eternal bottom line, I have tried to demonstrate that there are many answers to the many questions. Our modern quest for the smoking gun that produced the Star of Bethlehem ought to share the stage with other queries about it, just as our contemporary insistence that the only real universe is a universe of orbits is challenged by other points of view capable of yielding other kinds of truths. How else can we broaden our understanding of nature across culture? I find exploring borderlands of culture collision fascinating because boundaries are places where things change. And I am intrigued by the process of change; why and under what conditions change happens.

But change can be painful. Often it involves a process not unlike what Plato's cave dweller experienced when he was exposed to a dazzling new kind of light he hadn't seen before. Out of conflict often emerges a new common sense. We have seen it happen in the far-reaching case of evolution applied both to life and to matter in the universe as well as in conflicts among the disciplines (especially in Chapter 4 on the Dinosaurs and the Dogon).

Yet in the subterranean turf far below the battlefield we also find a kind of transcultural unity. Symbols, numbers, and taxonomies seem to unify the human quest to understand nature. The fourfold nature of the world intrudes upon us; the Maya on one side of the globe saw basically the same sky beheld by the Greeks; we all seek ways to represent space, to express time. The more hierarchically organized the society, the more time's expression seems to become a struggle to domesticate it and the more abstract and disconnected from mundane life the map that represents space becomes. Like the Greek orbital model of the universe, our maps and our clocks separate us from—rather than draw us into—experiencing time and space.

In unity there is diversity and out of diversity emerge more questions. The Star of Bethlehem, the Dogon diagram in the sand, the Inca *ceque* system, ancient Mexico's Fejéváry One diagram, the enigmatic ⊕. They are all symbols; and although I've tried to show how terribly important these symbols were to the people who contrived them, my external view of certain aspects of their internal reality makes it impossible for me to say that I know what those who made those symbols experienced, when they used them, who drew them, or who worked with them.

"What is true for the emotions may also be true for the intellect," writes cognitive psychologist Steven Pinker.[2] There's a mismatch between the purposes behind the evolution of our mental faculties and the way we use them today, such as when we gorge in anticipation of the famine that never comes and make babies we don't want. Such is the case when we think about the world around us, he goes on, for "[o]ur minds keep us in touch with aspects of reality—such as objects, animals, and people—that our ancestors dealt with for millions of years. But as science and technology open up new and hidden worlds, our untutored intuitions may find themselves at sea."[3]

In the last chapter, I cast myself in the role of the stumbling cave dweller. I discovered that the more I learned about a symbol, the more I seemed compelled to stray from the answer to its mystery that my "untutored intuition" tried to force upon me. I will admit that my intuition is partly based on disciplinary egotism—an innate desire to cast my own scientifically minded shadow over those who have gone before me. Perhaps thinking they were like me makes me feel more secure; maybe it validates the correctness of my world view. It took me more than a decade to admit that the astronomer's answer to the mystery of the ⊕ just doesn't satisfy. Sure, somebody in my kind of profession must have been in on the ground floor, for when we deal with a divine cosmos we all need to manage our symbolic apparatus correctly if we expect it to work. I think the people of ancient Teotihuácan shared my love of numbers as well. After all, getting the numbers right is part of the divination process.

Following the data where it leads—reasoning from the part to the whole, from the particular to the general—is what scientists call induction. It has always seemed to be the way I have proceeded. In my "confessional" chapter—call it my "tutored intuition" chapter—I tried to show by example how induction often can lead to blind alleys, especially when you ignore vital information that already may be out there. My more than thirty-year love affair with the ⊕ symbol—and the realization that to understand it, I must follow the facts no matter where they may lead—pointed to some truths about the many manifestations of that symbol that I never could have imagined at the outset. I began my journey in the halls of science and ended up in the temple of religion. Still, I'm not satisfied that I have the whole picture. Forever imprisoned by my own cognitive forms, I will never really know completely and definitively what that carving on the stone meant to them.

An eminent scholar who spent most of his life studying Maya hieroglyphs once remarked that we have about as much chance of penetrating the space-time mentality of the Maya people as does an atheist trying to understand the ecstasy of Saint Francis. There are so many conceivable ways of integrating all the impressions and images rained down upon the human senses. So here's the lesson: beware of falling into the trap of forcing your way of thinking on somebody else's symbol. The message contained in it may not have been intended for you.

I opened this narrative about ways of seeking truth in nature with a painting at an inn in my small village. I close it with one that hangs in my living room—a cherished four-by-five-foot abstract painting created in the 1960s by my friend and colleague, the artist Eric Ryan (Figure 36). It shows what looks like the prow of a green fishing boat emerging from mid-canvas amidst a confusion of blues, blacks, and whites. Over the boat lies a series of vertical black lines. At

36. The painting Considering Hellas in Eire *by Eric Ryan that hangs in my living room. (Photo by John Hubbard)*

the lower left a ten-inch wooden model of a red boat is pasted on the canvas in profile. The mysterious outline of a faceless man clad in a long, brown coat covers the whole of the right side of the canvas. Printed across the bottom in two-inch-high letters are the words *Considering Hellas in Eire.* Above floats the word *failte* ("welcome" in Gaelic) in smaller print.

Over the years visitors who have come to the house have gazed and gawked at our Ryan, each seeking the artist's intent. My father, often assisted by his afternoon cocktail, has offered several different analyses: "He's an Irishman longing to go on a fishing trip"; "An ocean disaster killed his father—that's the old man's ghost on the right"; "The black rain shows how gloomy he is"; "It's about a man who thinks he's from a Greek background"; or "He must have been a guy who *really* loved Greece." One of my relatives once remarked as he pointed to the blank face of the man outlined in brown contours: "How come your pal didn't finish his painting?"

Ryan was a dear friend who died young and tragically thirty-five years ago at the age of forty—not from a shipwreck but from a virulent flu epidemic. His lungs had been weakened from the bends when he was brought up too quickly while serving as a diver on one of Jacques Cousteau's expeditions exploring undersea artifacts off the Aegean coast of Turkey near the ruins of Bodrum. I haven't tried to push my privileged information about Ryan any further to try to decode the painting (which happens to be one of a triptych displaying similar abstract Aegean maritime scenes). Let's just say that it is safe to say he was an Irish guy who loved the Aegean. I want my Ryan to remain a mystery because I want my eye to continue to be attracted to it in hopes of finding something in it I hadn't known before. I seek no ultimate resolution, no bottom line to the myth behind why Ryan made it look the way it does. But I want to invite visitors into my cave so that I can learn their points of view—for that teaches me as much about them as it does about the painting.

NOTES

1. Plato, *The Republic,* ed. Benjamin Jowett (New York: Dover, 2000).

2. Steven Pinker, *The Blank Slate: The Modern Denial of Human Nature* (New York: Viking, 2002), 219.

3. Ibid.

FURTHER READING

CHAPTER 1

There are hosts of books and articles on the Star of Bethlehem. Most of the contemporary ones dwell, as one might expect, on attempts to identify the phenomenon. Among these are the books by Michael Molnar and David Hughes. To gain a more balanced view, readers should consult the work of Kim Paffenroth and my own piece in *Archaeology*, which raises broader questions.

Aveni, Anthony. "The Star of Bethlehem," *Archaeology* (November–December 1998): 35–42.

Buber, Martin. "The Wonder on the Sea." In Martin Buber, *Moses: The Revelation and the Covenant,* 74–79. New York: Harper, 1958.

Hughes, David. *The Star of Bethlehem: An Astronomers' Confirmation.* New York: Walker, 1979.
Molnar, Michael. *The Star of Bethlehem: The Legacy of the Magi.* New Brunswick, NJ: Rutgers University Press, 1999.
Paffenroth. Kim. 1993 "The Star of Bethlehem Casts Light on Its Modern Interpreters," *Quarterly Journal of the Royal Astronomical Society* 34 (1993): 449–460.

CHAPTER 2

Stephen Toulmin and June Goodfield's *The Fabric of the Heavens* gives a good introduction to geometry and orbits. Derek Price's *Science Since Babylon* offers some of the philosophical background behind the Greek way of thinking about the heavens, as does D. R. Dicks's *Early Greek Astronomy to Aristotle.* J. P. Vernant places Greek science in a social context; his ideas, although not universally agreed upon, have made an impact on my views. In *Conversing with the Planets* I deal specifically with the subject of astrology as a driving force in the pursuit of both Old and New World concepts regarding the structure of the heavens. I also discuss the great divide that developed between them during the Renaissance.

Aveni, Anthony. *Conversing with the Planets: How Science and Myth Invented the Cosmos.* Boulder: University Press of Colorado, 2002.
Dicks, D. R. *Early Greek Astronomy to Aristotle.* Bristol: Thames and Hudson, 1970.
Duhem, Pierre. *To Save the Phenomena: An Essay on the Idea of Physical Theory from Plato to Galileo.* Chicago: University of Chicago Press, 1985.
Price, Derek de Solla. *Science Since Babylon.* New Haven: Yale University Press, 1975.
Toulmin, Stephen, and June Goodfield. *The Discovery of Time.* Chicago: University of Chicago Press, 1965.
———. *The Fabric of the Heavens: The Development of Astronomy and Dynamics.* New York: Harper, 1961.
Vernant, J. P. *Myth and Thought Among the Greeks.* London: Routledge, 1983.

CHAPTER 3

Stephen Jay Gould's books, among them his 2002 collection of essays, give a thorough yet accessible account of Darwin's view of life, and Michael Behe's controversial *Darwin's Black Box* explains the idea behind intelligent design. Cosmic evolution is clearly expounded in Lee Smolin's 1997 book as well as in an equally accessible text by Martin Rees written that same year.

Behe, Michael. *Darwin's Black Box: The Biochemical Challenge to Evolution*. New York: Free Press, 1996.

Bowler, Peter. *Evolution, The History of an Idea*, 3rd ed. Berkeley: University of California Press, 2003.

Gould, Stephen Jay. *I Have Landed: The End of a Beginning of Natural History*. New York: Harmony, 2002.

Numbers, Ronald. *Creation by Natural Law: Laplace's Nebular Hypothesis in American Thought*. Seattle: University of Washington Press, 1977.

Rees, Martin. *Before the Beginning: Our Universe and Others*. Reading, MA: Helix/Perseus, 1997.

Smolin, Lee. *The Life of the Cosmos*. New York: Oxford, 1997.

CHAPTER 4

Broad in its extent, this chapter was based on my readings from a number of journals. The most accessible supplementary reading on various components of the chapter can be found in Kenneth Brecher's edited work (for the Dogon controversy), Charles Officer and Jake Page (for the dinosaur controversy), and in Joe Burchfield's book on Lord Kelvin (for the age of the earth controversy).

Brecher, Kenneth. "Sirius Enigmas." In *Astronomy of the Ancients*, ed. Kenneth Brecher and Michael Feirtag. Cambridge: MIT Press, 1979.

Burchfield, Joe. *Lord Kelvin and the Age of the Earth*. New York: Science History Publications, 1975.

Officer, Charles, and Jake Page. *The Great Dinosaur Extinction Controversy*. Reading MA: Addison Wesley, 1996.

Powell, James. *Night Comes to the Cretaceous: Dinosaur Extinction and the Transformation of Modern Geology*. New York: Freeman, 1998.

Van Sertima, Ivan, ed. *Blacks in Science: Ancient and Modern*. New Brunswick: Transaction, 1984.

CHAPTER 5

Annemarie Schimmel's *The Mystery of Numbers* is a fascinating introduction to the human preoccupation with numeracy. Specifically, I can recommend Trudy Griffin-Pierce's book on the Navajo *hooghan* (*hogan*), which offers a treasure trove on the role of the number four in Native American cosmology. Other primary sources are listed below.

Aaboe, Asger. "Scientific Astronomy in Antiquity." In *The Place of Astronomy in the Ancient World: A Joint Symposium of the Royal Society and the British Academy*, ed. F. R. Hodson. London: Oxford University Press for the British Academy, 1974.
Griffin-Pierce, Trudy. *Earth Is My Mother, Sky Is My Father: Space, Time, and Astronomy in Navajo Sandpainting.* Albuquerque: University of New Mexico Press, 1992.
Heidel, Alexander. *The Babylonian Genesis: The Story of Creation.* Chicago: University of Chicago Press, 1942.
Schimmel, Annemarie. *The Mystery of Numbers.* Oxford: Oxford University Press, 1993.
Tedlock, Dennis. *Popol Vuh, the Definitive Edition of the Mayan Book of the Dawn of Life and the Glories of Gods and Kings.* New York: Simon and Schuster, 1985.

CHAPTER 6

Historians Alfred Crosby and Jacques Le Goff have written lucid accounts of the impact of clock time upon medieval culture. On the Templo Mayor see Eduardo Matos Moctezuma's *The Great Temple of the Aztecs* and on the Aztecs in general see the work of Richard Townsend. My *Empires of Time* contains a more lengthy synthesis of the topics discussed in this chapter.

Aveni, Anthony. *Empires of Time: Calendar, Clocks and Cultures.* Boulder: University Press of Colorado, 2002.
Crosby, Alfred. *The Measure of Reality: Quantification and Western Society, 1250–1600.* Cambridge: Cambridge University Press, 1997.
Le Goff, Jacques. *Time, Work and Culture in the Middle Ages.* Chicago: University of Chicago Press, 1980.
Lippincott, Kristen, ed. *The Story of Time.* London: Merrill Holberton and Natural Maritime Museum, 2000.
Matos Moctezuma, Eduardo. *The Great Temple of the Aztecs.* London: Thames and Hudson, 1978.
Townsend, Richard. *The Aztecs.* London: Thames and Hudson, 1992.

CHAPTER 7

Without doubt the best source on maps is the multivolume compendium edited by John Harley and David Woodward. On non-Western representational systems discussed in this chapter, see David Lewis on Polynesia and my piece in the book edited by Merald Wrolstad and Dennis Fisher, which discusses both stick charts and the *ceque* system.

Aveni, Anthony. "Non Western Notational Frameworks and the Role of Anthropology in Our Understanding of Literature." In *Toward a New Understanding of Literacy,* ed. Merald Wrolstad and Dennis Fisher, 252–280. New York: Praeger, 1986.

Harley, John B., and David Woodward. *The History of Cartography,* vol. 2, book 2. Chicago: University of Chicago Press, 1984.

Lewis, David. *We the Navigators: The Ancient Art of Landfinding in the Pacific.* Honolulu: University Press of Hawaii, 1975.

Turnbull, David. *Maps Are Territories: Science Is an Atlas.* Chicago: University of Chicago Press, 1993.

CHAPTER 8

One of the reasons I decided to write this chapter is that I was surprised to find that there is little in the way of an accessible, general synthesis of non-Western taxonomies. Brent Berlin, Dennis Breedlove, and Peter Raven offer the best discussion of Maya plant taxonomy. Keith Thomas's *Man and the Natural World* makes for excellent reading on the subject of pre-Linnaean Western taxonomies, and Heinz Goerke's *Linnaeus Systema Naturae* makes the transition. I also found Harriet Ritvo's fascinating book on taxonomic oddities very helpful.

Berlin, Brent, Dennis Breedlove, and Peter Raven. *Principles of Tzeltal Plant Classification.* New York: Academic Press, 1974.

Goerke, Heinz. *Linnaeus Systema Naturae.* New York: Scribners, 1973 [1735].

Ritvo, Harriet. *The Platypus and the Mermaid and Other Figments of the Classifying Imagination.* Cambridge: Harvard University Press, 1997.

Thomas, Keith. *Man and the Natural World.* New York: Pantheon, 1983.

CHAPTER 9

I am responsible for a good bit of the work on the pecked cross-circles. Readers might consult my piece in *Science,* my most recent discussion of this pan-Mesoamerican symbol in the *Journal for the History of Astronomy,* and my *Skywatchers* for references in the journals to significant work by others should they wish to engage this subject at a deeper level.

Aveni, Anthony. "Observations on the Pecked Designs and Other Figures Carved on the South Platform of the Pyramid of the Sun at Teotihuacan." *Journal for the History of Astronomy* 36:1 (2005): 30–47.

———. *Skywatchers: A Revised Updated Version of Skywatchers of Ancient Mexico.* Austin: University of Texas Press, 2001.

———, Horst Hartung, and Beth Buckingham. "Pecked Cross Symbol in Ancient Mesoamerica" *Science* 202 (1978): 267–279.

INDEX

Page numbers in italics indicate illustrations